TECHNICAL DRAWING

Dennis Maguire, C.Eng., M.I.Mech.E., Mem.ASME, MIED, is Senior Lecturer in Engineering Drawing and Design in the Mechanical and Production Engineering Department at Southall College of Technology. He is also chief Examiner of the City and Guilds of London Institute in Engineering Drawing for the general course in Engineering. He has served for the past ten years as a member of council in the Institution of Engineering Designers and is at present Chairman of the Institution's Design Promotion Committee.

Colin Simmons, Mem.ASME, FIED, has spent many years actively engaged in the draughting field and is now the Standards Engineer at Lucas CAV Ltd. Engineering experience includes the design of aero engines, high speed diesel engines and fuel injection equipment. He is also Chairman of the British Standards Draughting Committee on general principles and a member of associated committees dealing with dimensioning and geometrical tolerancing. He is the United Kingdom delegate on ISO committees TC10/SC1 and SC5 which are the international counterparts.

TEACH YOURSELF BOOKS

TECHNICAL DRAWING

D. E. Maguire and C. H. Simmons

TEACH YOURSELF BOOKS
Hodder and Stoughton

First Impression 1980
Third Impression 1983

British Library C.I.P.

Maguire, Dennis Eric
Teach yourself technical drawing. –
(Teach yourself books).
1. Engineering drawings
2. Mechanical drawing
I. Title II. Series
604′.2′.4 T353

ISBN 0–340–24790–8

Filmset by Northumberland Press Ltd, Gateshead, Tyne and Wear
Printed and bound in Great Britain for
Hodder and Stoughton Educational,
a division of Hodder and Stoughton Ltd,
Mill Road, Dunton Green, Sevenoaks, Kent
by Richard Clay (The Chaucer Press) Ltd, Bungay, Suffolk

Contents

Acknowledgements

The authors express their special thanks for assistance and permission to publish material from the following;

British Standards Institution, 2 Park Street, London, W1A 2BS, from whom complete copies of the standards quoted may be obtained.

British Thornton Limited, Wythenshawe, Manchester.

Letraset Limited.

Associated Lancashire Schools Examining Board.

City and Guilds of London Institute.

East Anglian Examinations Board.

Middlesex Regional Examining Board.

North Western Secondary School Examination Board.

Audrey and Beryl, our wives and typists.

Kate, our kind friend and patient tracer.

The Editor and Staff of Teach Yourself Books.

Introduction

Technical drawing has gradually developed into an efficient international form of graphical communication between designers, craftsmen and consumers.

Clear unambiguous instructions must be conveyed by drawings and National Standards continually attempt to improve minor details in order to achieve worldwide acceptability. The current British Standard is BS308: 1972, an abridged edition for use by schools and colleges is PD7308: 1980. This book is written to conform with these Standards.

It is not necessary for all technicians to be skilled draughtsmen but in order to appreciate fully all of the information conveyed by the drawing it is most important to understand the reason for the inclusion of each of the lines and notes.

The most advanced drawings may include hundreds of dimensioned details but each represents basic information relating to shape, location, structure and finish of individual but related parts. Large drawings are prepared and developed by arranging together a variety of features which are geometric in form and the draughtsman must be familiar with the associated principles of geometry.

In this book we have presented some necessary geometry with technical applications to enable you to reach a reasonable standard of competence as a draughtsman. We hope you will redraw the dimensioned illustrations as technical exercises to gain first-hand experience and then attempt the examination style problems which follow. All solutions are provided so that you can teach yourself by example. Some of the applications use very simple components and these are quite deliberately included because we are trying to establish basic elementary principles. We also recommend the use of models as comprehension aids for many of the examples of solid geometry and orthographic projection. Models can be cut very easily from the rigid polystyrene type of materials, used for packing, with a sharp penknife.

We sincerely hope that the text which follows will enable you to obtain a sound basic understanding of technical drawing and will also stimulate a desire for further study.

1

Technical Drawing Equipment; Linework; Drawing Standards and Draughting Conventions

Draughting Equipment

Some basic equipment is necessary in order to teach yourself drawing. This should be reliable and accurate as precision draughting equipment is required to produce good linework. The following brief notes are intended to help you select a few essential items.

The drawing board and T-square

All of the illustrations in this book were prepared on a drawing board which accommodated an A2 size of drawing sheet. The A2 size of paper is 420 × 594 mm.

A drawing board should be manufactured from well seasoned wood and fitted on the reverse side with battens to prevent warping. The working edge at the left hand side may also be fitted with a hard wood insert or a specially prepared metal strip which protects the edge and permits the T-square to slide up and down with ease.

When not in use, always store the board away from direct sunlight to prevent warping and also ensure that the board is not subjected to extremes of heat and humidity.

The T-square is used for drawing horizontal parallel lines across the paper and this is often made from wood. Treat the edge of the T-square with special care so that it is always smooth. Some squares are available with transparent strips along the working edge and this is an advantageous feature since the draughtsman can view part of the drawing beneath the T-square. When using a T-square, always ensure that the stock, which is the vertical part of the square, is held firmly against the side of the drawing board.

Students are generally not allowed to use drawing pins to secure the drawing paper or film to the drawing board since pins ruin the surface. Draughting tape is available to keep the paper secure at the corners, or alternatively, you can use drawing paper clips. These are inexpensive and only two are required to secure the top edge of the paper to the board.

It is a common practice to cover the drawing board surface with a backing sheet of paper before use and this helps to maintain the board in first class condition.

Parallel slide attachments are available which enable the draughtsman to draw parallel lines with ease. The drawing board is fitted with a stand and a modified T-square is given a parallel motion by a simple pulley system.

In recent years, washable, lightweight boards have also been manufactured in high impact plastics materials. These boards are fitted with a sliding set square which is transparent and often engraved with a millimetre scale. Locking devices hold the paper against the board. These boards can be inspected at any large drawing equipment dealer and can be recommended.

Set squares

Set squares are often known as triangles because of their shape. An adjustable set square is a desirable item of equipment which is fitted with a protractor and a rotating arm. The draughtsman can conveniently lay out lines at any angle with an adjustable set square. Many sizes are available but an arm length of 250 mm would be suitable for most drawing purposes. Alternatively, set squares with fixed angles of 45° and 60° will be necessary, plus a protractor to mark out other angles. All of this equipment is made from a tough transparent plastics material so that the drawing can be seen beneath the square.

Compasses

A compass is required to draw circles and the draughtsman usually possesses a large and small compass. The compass should be rigid so that the same

Fig. 1.1 Typical draughting equipment.

1. Drawing board
2. 'T' Square
3. 45° set square
4. 60° set square
5. Protractor
6. Small spring compass/divider
7. Large spring compass/divider
8. Adjustable triangle
9. Parallel slide motion drawing board with adjustable stand

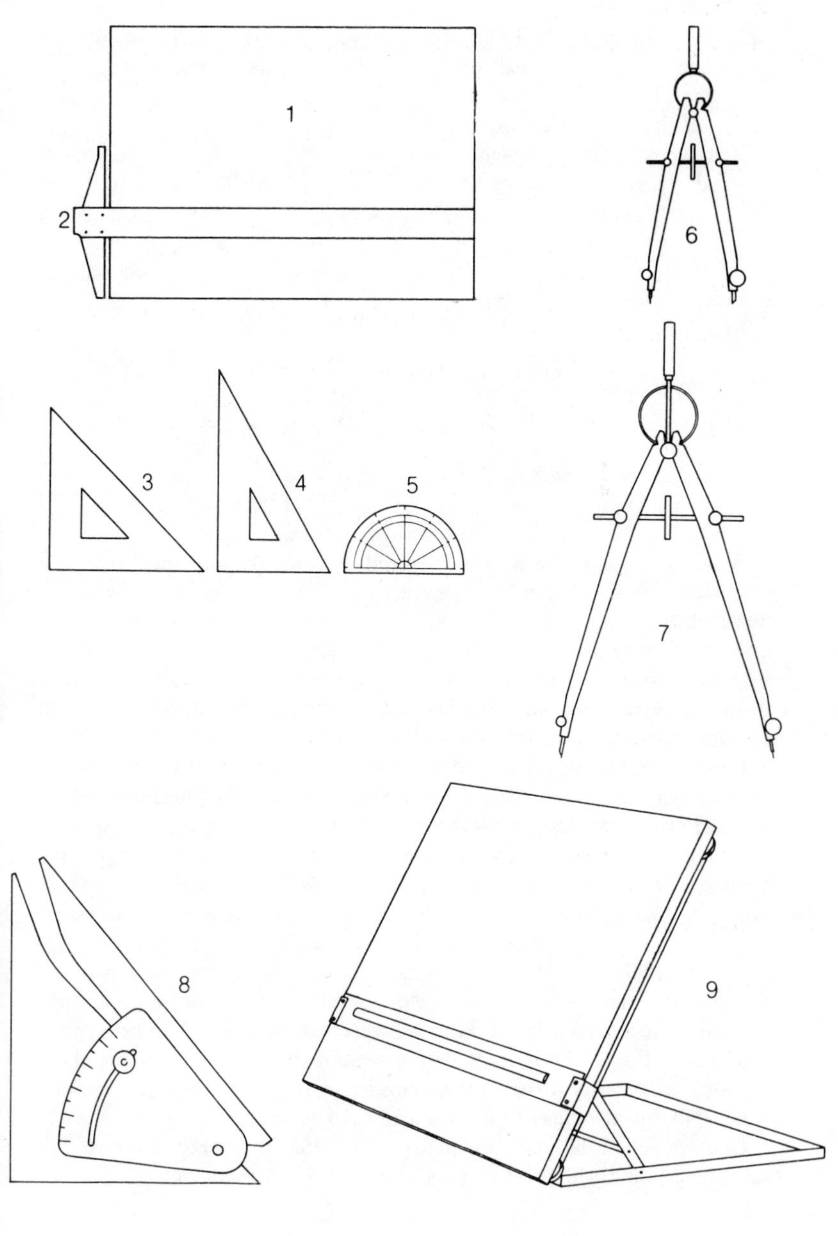
1
2
3
4
5
6
7
8
9

radius is maintained while a clean circular arc is drawn, and, in the authors' experience, this is most easily obtained if the instrument has a centre adjusting screw. The compass lead should be kept sharp by rubbing on a sandpaper block. Hold the compass so that the legs are in line and at right angles to the block. After sharpening, adjust the length of the pin so that it protrudes just slightly longer than the lead. When drawing a circle, the compass head should be lightly held between the index finger and the thumb and only with a little pressure glide the lead around the centrepoint. With practice, most students get the feel of the gentle action which is necessary to draw circles and arcs of consistent line density.

Attachments are available to permit ink drawing.

Drawing pencils

Students of technical drawing usually draw on paper surfaces but these vary in thickness and texture and a little experimentation is necessary to find the degrees of pencil hardness which give satisfactory construction lines and finished linework. Good quality pencils of hardness HB, H, 2H and 3H will generally be found suitable for most surfaces.

When drawing a line with a fine point of a newly sharpened pencil, a consistent line width can be achieved by rotating the pencil slowly as the line is drawn along the straight edge. If the pencil lead wears on one side only, then the pencil will produce lines of varying density and width depending on the position in which it is held. A pencil is correctly held about 2–3 cm from the point. Note that the hexagonal shape of the wood casing will provide a sure hold without having to exert a tight grip.

Clutch type pencils are available at most good drawing office equipment suppliers and one type takes a 2 mm diameter lead. The extent to which the lead protrudes from the casing is controlled by a push button operating a clutch mechanism. An inexpensive lead pointer is used to sharpen the lead or as an alternative, a chisel point can be formed with a sandpaper block. A further range of clutch pencils are also on sale which do not need sharpening. One type of pencil holds round leads but separate pencils are required for different lead thicknesses. Pencils with lead thicknesses of 0·7 mm and 0·35 mm will give satisfactory results and contrasting linework. The leads are available in the different degrees of hardness. Another type of clutch pencil uses ribbon leads with a rectangular cross section and these are also satisfactory for producing consistent widths of lines but separate pencils are again required for each lead thickness.

One advantage of using a clutch pencil is that the pencil is always the same length and will always have the same 'feel' in the hand.

A well balanced pen will assist a writer with handwriting and good quality pencils are invaluable to the draughtsman for the production of sound linework.

Erasers

A good block-shaped vinyl or rubber eraser can be used for both pin point erasure and for cleaning large areas of a drawing. With a vinyl eraser, the debris curls into easily disposable particles which can be brushed or blown away and excessive hand pressure is not necessary or desirable.

An erasing shield is also a handy tool. These are inexpensive, made from stainless steel or a plastics material and various apertures on the surface allow erasure of drawing detail while protecting the surrounding linework.

Scale rules

The student will require a rule with divisions in millimetres and centimetres and the standard length of 30 cm is quite suitable. Later on, layout drawings of house plans or enlarged drawings of small parts, for example, may be undertaken and scale rules are available with suitably engraved scales to assist the draughtsman without his having to resort to mental arithmetic.

Basic draughting equipment

The equipment discussed will meet your needs while you are still in the early stages of technical drawing. A lot of equipment is not needed and the tendency is to provide students with too many items which are seldom used. However, you should inspect drawing instruments at specialist suppliers and if possible seek the advice of an experienced draughtsman. It must be stressed that technical drawing requires a degree of precision which cannot be obtained using the conventional equipment found in school geometry sets so some importance should be attached to the selection of reliable and accurate equipment.

Drawing Sheet Sizes

Drawing sheets in current use are sized according to the ISO 'A' series. The range of preferred sizes of the trimmed sheets selected from the main ISO 'A' series are tabulated below.

Designation	*Dimensions in millimetres*	*Surface area*
A0	841 × 1189	1 sq metre
A1	594 × 841	5000 sq cm
A2	420 × 594	2500 sq cm
A3	297 × 420	1250 sq cm
A4	210 × 297	625 sq cm

The drawing sheets are not square, the sides being in the ratio of $1{:}\sqrt{2}$.
Reference to the table shows that:

the A1 size is half of the area of the A0 size
the A2 size is half of the area of the A1 size
the A3 size is half of the area of the A2 size
the A4 size is half of the area of the A3 size

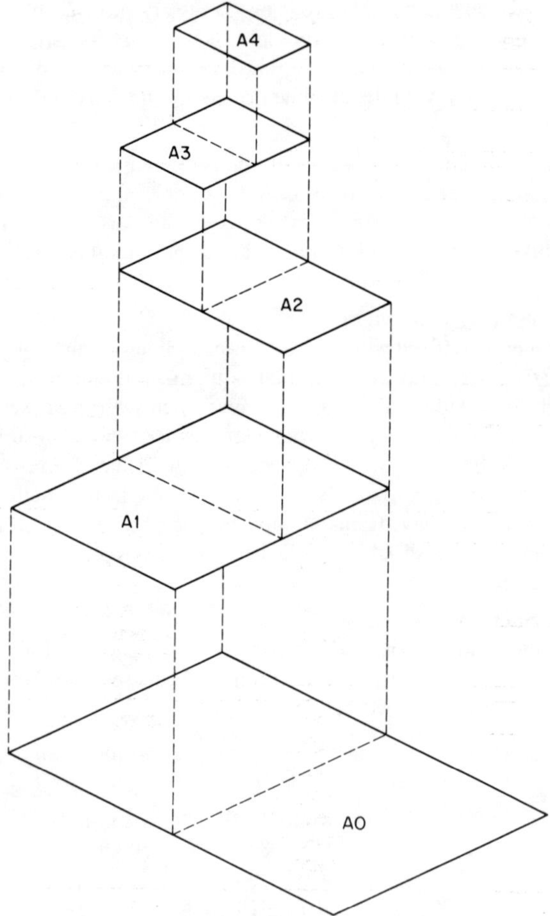

Fig. 1.2 Relationship of the 'A' sizes of drawing sheets.

The A0 size of one square metre lends itself to a paper weight unit and the various grades are expressed in terms of 'grams per square metre'.

It is customary for drawing sheets to incorporate a frame enclosing the drawing area. The width of the frame recommended is 10 mm–20 mm, according to the paper size.

Type of line	Description of line	Application
	Thick, continuous	Visible outlines and edges
	Thin, continuous	Dimension and leader lines Projection lines Hatching Outlines of adjacent parts Outlines of revolved sections
	Thin, continuous irregular	Limits of partial views or sections when the line is not an axis
	Thin, short dashes	Hidden outlines and edges
	Thin, chain	Centre lines Extreme positions of movable parts
	Chain (thick at ends and at changes of direction, thin elsewhere)	Cutting planes

Table A Types of line

A title block is positioned at the bottom right hand corner of the frame. Basic information in the block includes the drawing title, scale, number, the draughtsman's name, the date of the drawing and the projection symbol where applicable. Projection symbols are explained later in this book.

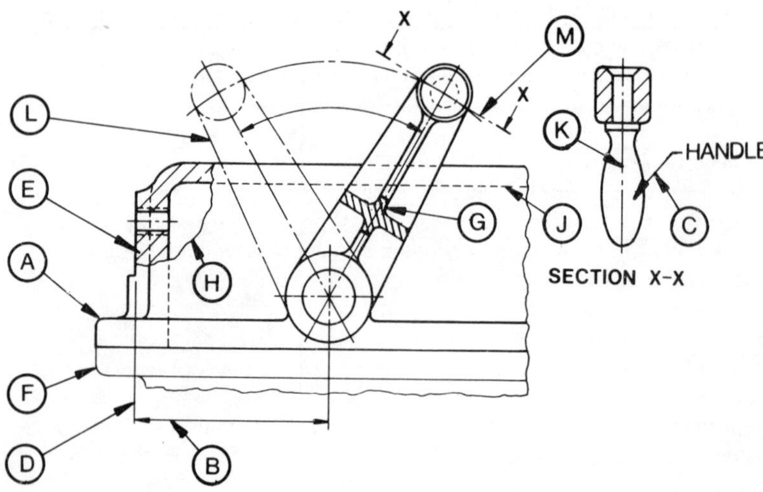

Fig. 1.3 Linework applications.

A. Visible outline
B. Dimension line
C. Leader line
D. Projection line
E. Hatching line
F. Outline of adjacent part
G. Revolved section outline
H. Limit of partial view
J. Hidden outline
K. Centre line
L. Extreme position of movable part
M. Cutting plane

Linework

Lines on any one drawing should all be in pencil, or all in black ink, uniformly black, dense and bold.

Two line thicknesses are required, a thin line and a thick line, approximately twice the thickness of the thin line. The Engineering Drawing Standard BS308:1972 recommends the line thicknesses to be 0·3 mm and 0·7 mm. All engineering drawing requirements can be obtained by using these line thicknesses and Table A shows the various types of lines, with their applications, which are freely used later in this book.

Linework density

The density of all lines should be such as to ensure clear reproduction when a print is taken from an original drawing. Lines are usually drawn in black ink or pencil and commercial grades of ink currently available are of sufficient density to give perfect reproduction. Graphite lead pencils produce lines which are made up of ridges of graphite flakes deposited on the drawing sheet as the pencil passes across. Pencil wear is inevitably uneven and the harder the pencil lead, then the harder it is for the surface of the drawing sheet to wear off a dense graphite deposit, which is necessary for good reproduction. Since there are various types and qualities of drawing materials available, you should experiment to find what suits you best.

Chain lines

The appearance of a sound drawing can be spoilt by carelessly produced dotted lines and centre lines. Try to observe the following simple rules;

1 Ensure that the dashes and spaces in a dotted line are of the same length throughout the line. A dash of about 3 mm followed by a space of 2 mm will produce a dotted line in good proportion.

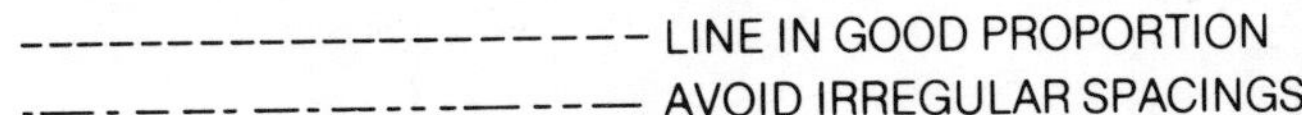

2 Where centrepoints are defined, then the chain lines should cross one another at a solid portion of the line and not at the spaces.

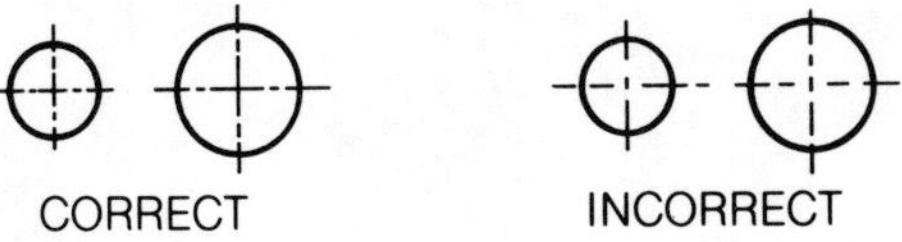

3 Centrelines should not extend through the spaces between views and they should also not terminate at another line on the drawing.

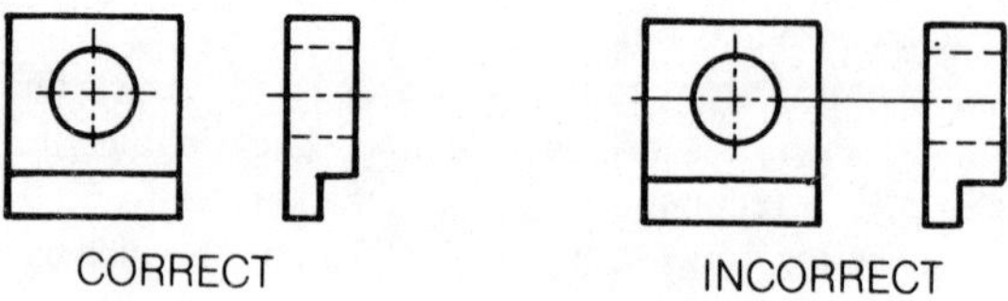

4 Where an angle is formed by chain lines, then the long dashes should intersect and define the angle.

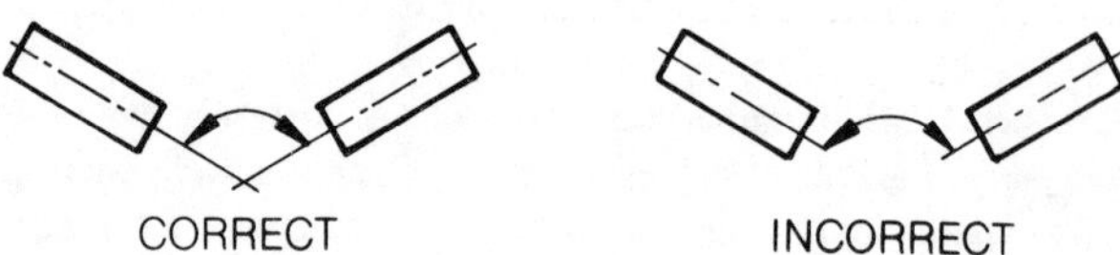

5 Generally, dotted lines which represent hidden detail must touch an outline without a break as shown below. Dotted lines also meet and cross each other and the junction should be arranged like a letter 'T' or 'X'.

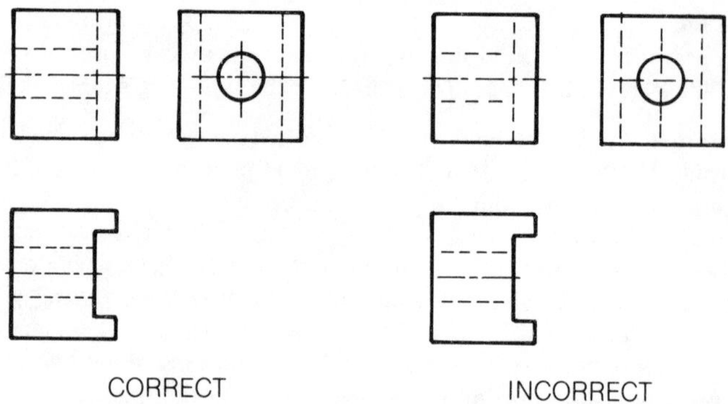

Lettering

Considerable care should be taken with lettering as lack of attention to small details can spoil the overall effect of a well prepared drawing. The styles reproduced below are provided as a guide in the British Standards. Note that vertical or sloping (italic) characters are acceptable, but that both types should not be used on the same drawing.

Capital letters are preferred to lower case letters since they are easier to read on reduced size drawing prints although lower case letters are used where they form part of a symbol or an abbreviation.

ABCDEFGHIJKLMN OPQRSTUVWXYZ	*ABCDEFGHIJKLMN* *OPQRSTUVWXYZ*
abcdefghijklmnopqrstuvwxyz	*abcdefghijklmnopqrstuvwxyz*
1234567890	*1234567890*
1234567890	*1234567890*

Fig. 1.4

Clarity and style are particularly important for the characters and they should be open in form and devoid of serifs and other embellishments. The spacing around each character is important to ensure that 'filling in' will not occur during reproduction. All strokes should be black and of consistent density.

Notes should not be underlined since this impairs legibility but where information is tabulated, the letters or numerals should be kept well clear of the spacing lines.

Attention is drawn in the Standards to the size of letters and characters and the table below gives the recommendations for minimum sizes on particular drawing sheets.

Application	Drawing sheet size	Minimum character height
Drawing numbers, etc.	A0, A1, A2, and A3 A4.	7 mm 5 mm
Dimensions and notes	A0. A1, A2, A3 and A4	3.5 mm 2.5 mm

Table B

The above dimensions refer to capital letters, but if lower case letters are used, then they should be drawn approximately 0·6 times the height of the capital letter. The stroke thickness should be drawn at approximately 0·1 times the character height and the clear space between characters at about 0·7 mm for 2·5 mm capitals and the other sizes in proportion.

The spaces between lines of lettering should be consistent and preferably not less than half of the character height.

Stencils are available for characters and letters in a variety of styles and sizes.

The illustrations in this book were prepared in ink on tracing paper and the notes and dimensions added using the 'Letraset' system. Fig. 1.5 shows a typical sheet of 'Letraset' letters and numbers and these are applied to the drawing surface by applying a slight pressure on the top surface of the sheet and the letter detaches itself from the back of the sheet and adheres instantly to the drawing. In the event of a mistake being made, the letter can easily be erased with no damage to the drawing surface. Many different styles of characters are available and the method is a popular one with draughtsmen.

Conventional representation of common features

The complete delineation of commonly used components is very time consuming and conventions have been devised to avoid unnecessary wastage of time. In Fig. 1.6 typical examples of some features used in mechanical engineering are given and applications of the conventions will be found in

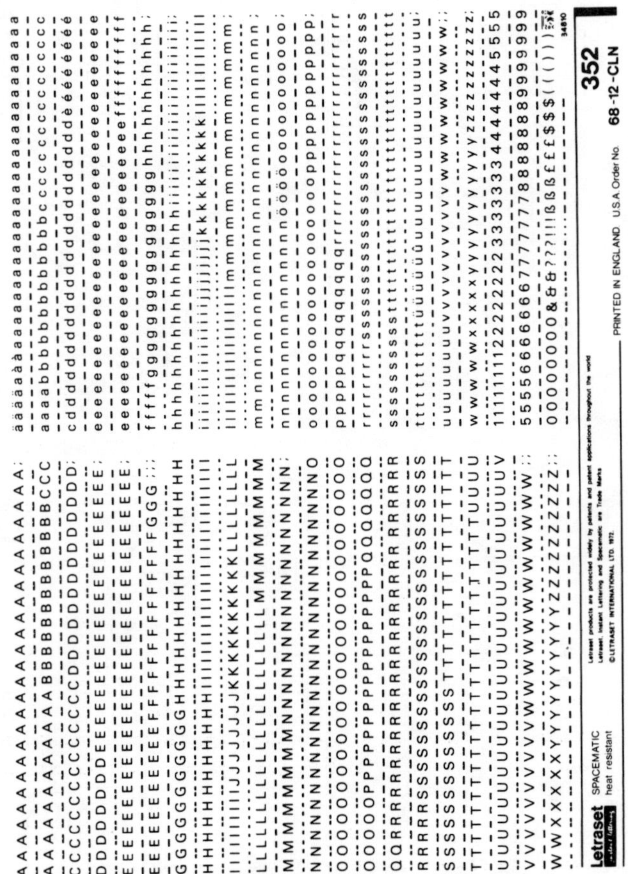

Fig. 1.5 Letraset instant lettering.

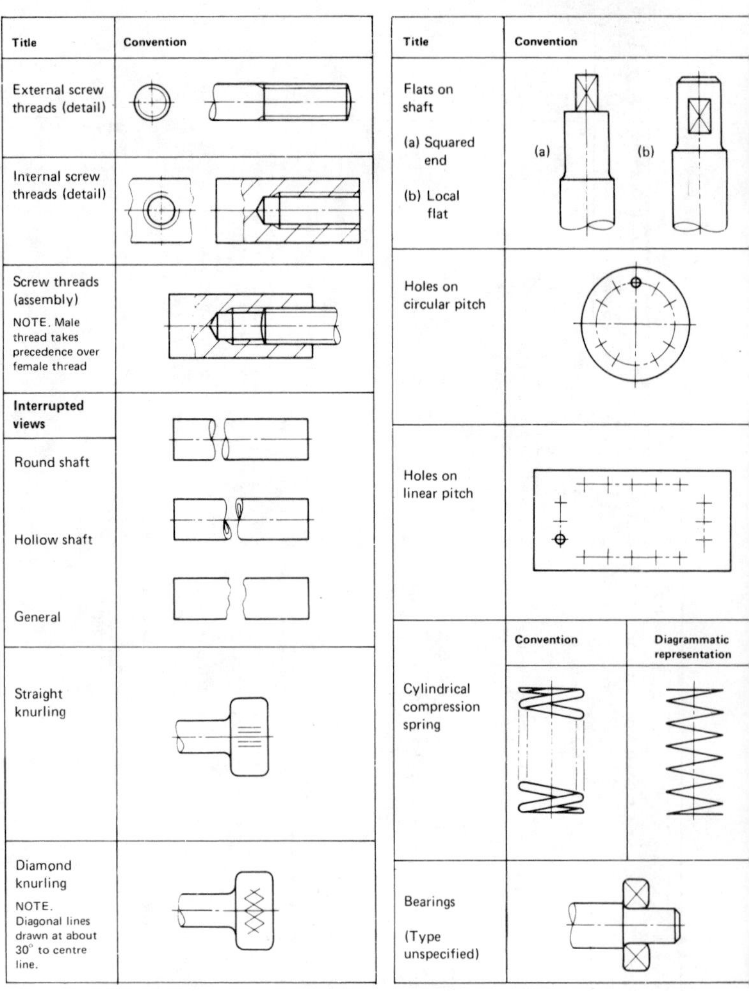

Fig. 1.6 Conventional representation of common features.

General engineering terms

Term	Abbreviation or symbol
Across flats	A/F
Assembly	ASSY
Centres	CRS
Centre line	L or CL
Chamfered	CHAM
Cheese head	CH HD
Countersunk	CSK
Countersunk head	CSK HD
Counterbore	C'BORE
Cylinder or cylindrical	CYL
Diameter (in a note)	DIA
Diameter (preceding a dimension)	Ø
Drawing	DRG
External	EXT
Figure	FIG.
Hexagon	HEX
Hexagon head	HEX HD
Hydraulic	HYD
Insulated or insulation	INSUL
Internal	INT
Left hand	LH
Long	LG
Material	MATL
Maximum	MAX
Minimum	MIN

Term	Abbreviation or symbol
Number	NO.
Pattern Number	PATT NO.
Pitch circle diameter	PCD
Pneumatic	PNEU
Radius (in a note)	RAD
Radius (preceding a dimension)	R
Required	REQD
Right hand	RH
Round head	RD HD
Screwed	SCR
Sheet	SH
Sketch	SK
Specification	SPEC
Spherical diameter (preceding a dimension)	SPHERE Ø
Spherical radius (preceding a dimension)	SPHERE R
Spotface	S'FACE
Square (in a note)	SQ
Square (preceding a dimension)	□
Standard	STD
Undercut	U'CUT
Volume	VOL
Weight	WT
Taper, on diameter or width	▷—

Table C

later chapters. This is only a small sample to draw the reader's attention to the use of drawing conventions.

Drawing abbreviations

In order to produce drawings quickly and efficiently, many lengthy terms have been abbreviated and standardised. The current British Standard, BS308, includes a list of abbreviations and symbols which are widely used in general engineering. The shortened list, shown on p. 17, Table C, gives many of the common items which are used freely in the drawing office and in this book.

2

Plane Geometry Required by the Draughtsman

Practical plane geometry

The ancient Egyptians are credited with one of the earliest known applications of practical geometry. They were faced with the problem of restoring boundaries which were removed due to flooding of the River Nile.

Scholars in ancient Greece developed the basis of our system of geometry and the word itself is derived from two Greek words: 'ge' meaning 'the earth', combined with 'metron', meaning 'measure'.

Geometry can be subdivided into two parts:

(*a*) Plane geometry which deals with lines and the construction of figures on a flat surface, often referred to as a two-dimensional geometry.

(*b*) Solid geometry which deals with three-dimensional solid objects.

Geometry is used by engineers, architects and artists for setting out their designs, in constructional work, in the interpretation of drawings and in inspection and checking, hence a thorough knowledge of geometrical principles is considered to be necessary.

In this chapter we have reproduced geometrical principles relating to the straight line, simple geometrical figures, and have concluded with geometry relating to the circle. An understanding of these principles is required to appreciate the applications which follow later in the book.

Angles

If two lines meet at a point, then an angle is said to be formed. In Fig. 2.1, a typical angle is illustrated and designated ABC. The point B is known as the vertex. We could, of course, describe the angle as CBA but it is customary to use letters in alphabetical order.

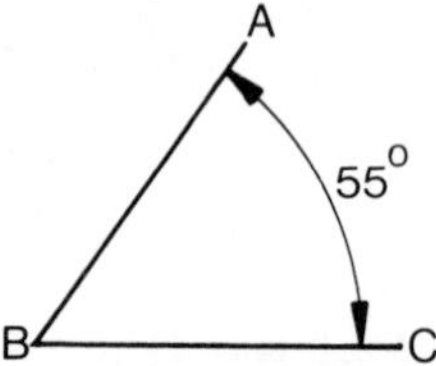

Fig. 2.1 Note the angle is dimensioned using an arc with its centre at the vertex B.

The unit for measuring an angle is the degree. One degree = 1°.

On the face of a clock, when a hand has moved through one complete revolution, the tip of the hand has traced out a circle and turned through 360°. If a circle is divided into two parts by drawing a line through the centre, we have formed two semicircles. By dividing a circle into four equal parts, we have formed four quadrants.

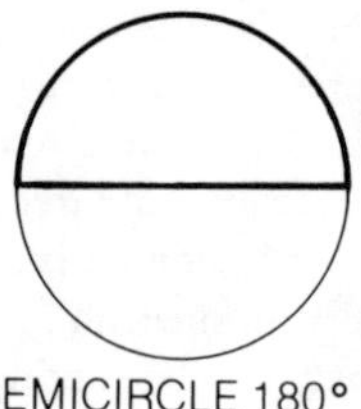

Fig. 2.2

Angles may also be named as follows:
An *acute angle* is less than 90°.
A *right angle* is 90°.
An *obtuse angle* is greater than a right angle but less than 180°.
A *reflex angle* is greater than 180°, but less than 360°.

The term 'complementary angles' is often used in geometry and this refers to two angles which add up to 90°, for example, 30° will be the complement

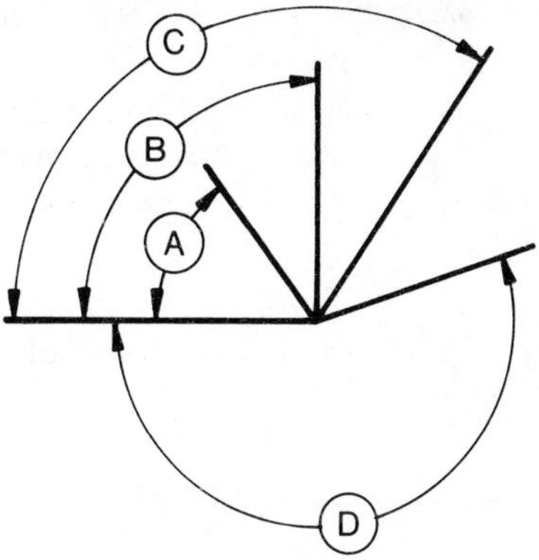

Fig. 2.3

A is an **acute** angle

B is a **right** angle

C is an **obtuse** angle

D is a **reflex** angle

of 60°. Similarly, the term 'supplementary angles' refers to two angles which add up to 180°, for example, 60° is the supplement of 120°.

The degree may be further subdivided into minutes and seconds as follows:

60 seconds = 1 minute and is written $60'' = 1'$
60 minutes = 1 degree and is written $60' = 1°$
360 degrees = 1 revolution and is written $360° = 1$ rev

To draw angles of 30°, 45°, 60° and 90°, the draughtsman uses set squares which are triangles made from a clear plastics material. Two types are in general use, one with angles of 30°, 60° and 90°, known as a 30° or a 60° set square, and the other with angles of 45°, 90° and 45° described as a 45° set square.

Angles may also be drawn using a protractor. The centrepoint of the protractor is positioned over the vertex of the required angle and the angle marked off against an engraved scale around the edge. Protractors are available in the shape of a complete circle subdivided into 360 degrees, or as a semicircle with 180 degrees.

An almost indispensible item of equipment for the draughtsman is an adjustable triangle which has a moving side fixed to a rotating protractor scale. This equipment is illustrated on p. 5.

All of the constructions which follow have applications in general draughtsmanship. Redraw each of the illustrations to gain experience in elementary geometry. Dimensions are not important.

To erect a perpendicular at any point P on given line AB (Fig. 2.4)
Draw a semicircle on AB with point P as centre and of any convenient

size. From X and Y, draw arcs of the same radii to intersect at C. Join C to P. APC = BPC = 90°.

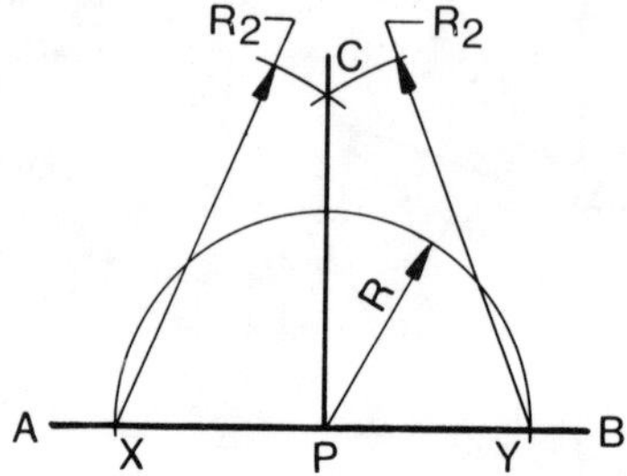

Fig. 2.4 *Note:* R is the abbreviation for radius and the symbol R_2 should be interpreted as a second radius of a different size.

To drop a perpendicular from any point P to a given line AB (Fig. 2.5)

Draw an arc from point P to intersect the given line at points X and Y. From X and Y, draw arcs of the same radii to intersect at point C. The line PC is perpendicular to AB.

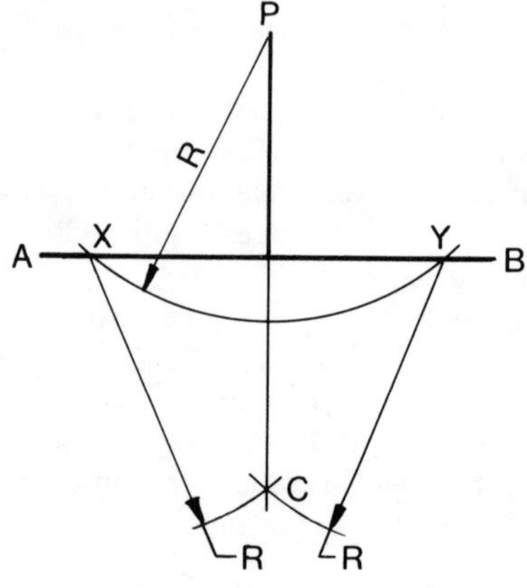

Fig. 2.5

To erect a perpendicular at any point P on a given line, where P is near to one end of the line (Fig. 2.6)

Draw part of a circle to pass through the required point P. The circle centre is shown as O, the circle intersects line AB at point X. Draw a line from X through centre O to intersect the circle again at point C. Draw line CP which is the required perpendicular.

This construction uses the fact that a triangle constructed in a semicircle with one side as diameter will be a right angled triangle. Draw alternative triangles as shown in Fig. 2.7 and check the angles at the circumference.

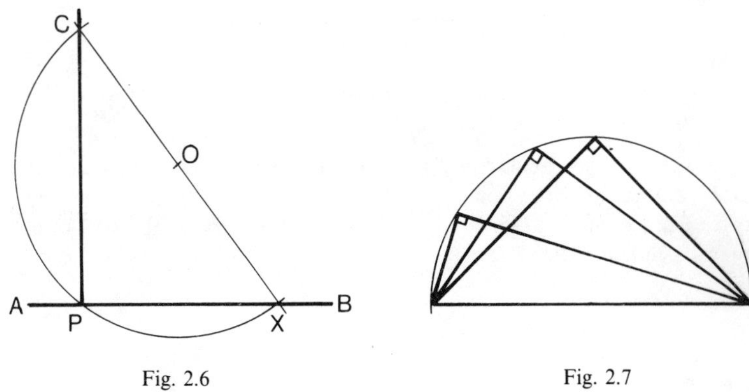

Fig. 2.6

Fig. 2.7

To bisect an angle (Fig. 2.8)

Draw any angle ABC. From B, draw an arc to intersect AB and CB at X and Y. From X and Y draw arcs using the same radius to intersect at point D. Join DB. Angle ABD = Angle CBD.

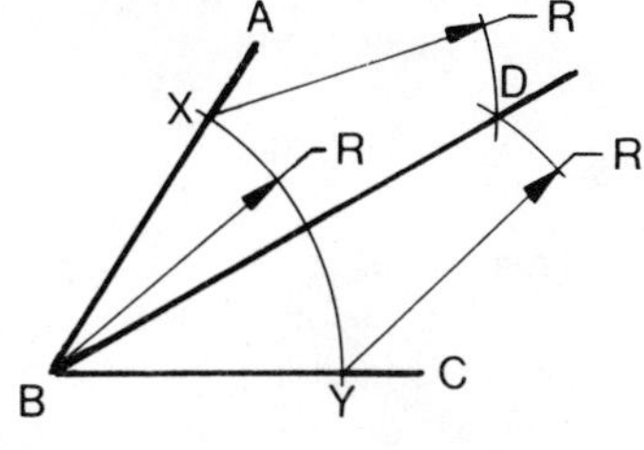

Fig. 2.8

To construct an angle of 60°

Draw any circle and it will be found that the radius can be stepped off around the circumference six times as shown in Fig. 2.9.

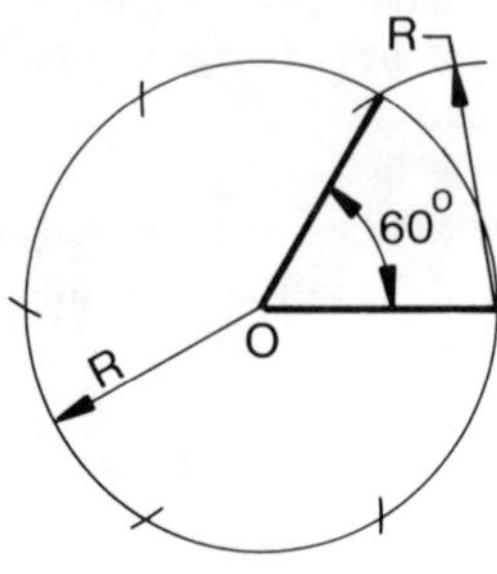

Fig. 2.9

To erect a perpendicular at point P on a given line using circular arcs

Draw any radius from point P as shown in Fig. 2.10. Step off the radius twice from point A to give points B and C. Bisect the angle BPC to give the perpendicular line PD.

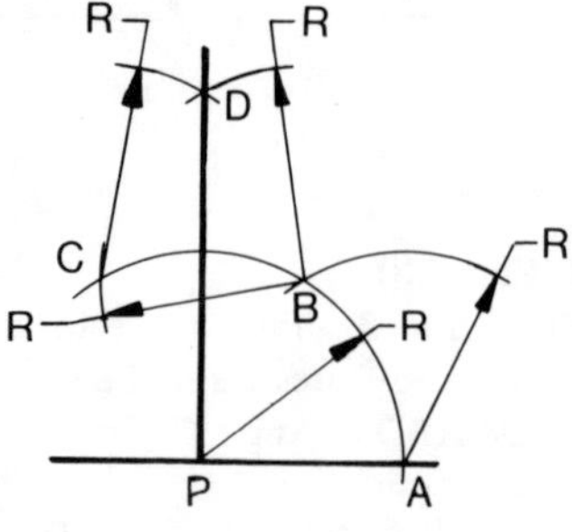

Fig. 2.10

Types of triangles

A *Scalene* triangle has three unequal angles. All sides have unequal lengths. The shortest side is positioned opposite to the smallest angle. The longest side is positioned opposite to the largest angle.

An *Obtuse* angled triangle has one angle greater than 90°.

An *Acute* angled triangle has all three angles of less than 90°.

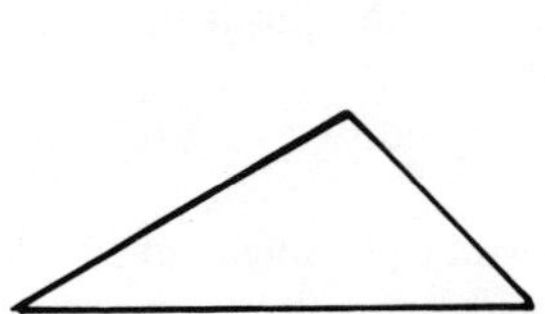

Fig. 2.11 Scalene and obtuse angled trangle.

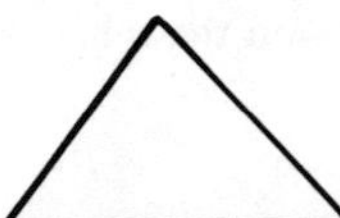

Fig. 2.12 Scalene and acute angled triangle.

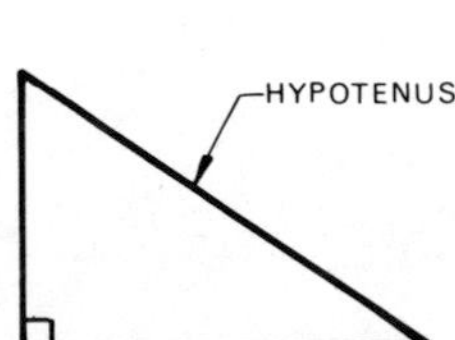

Fig. 2.13 Right angled triangle.

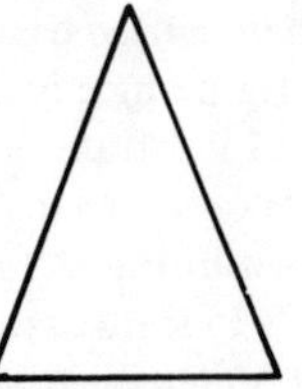

Fig. 2.14 Isosceles triangle.

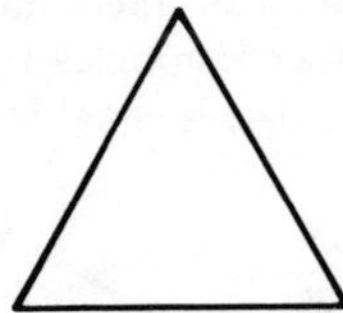

Fig. 2.15 Equilateral triangle.

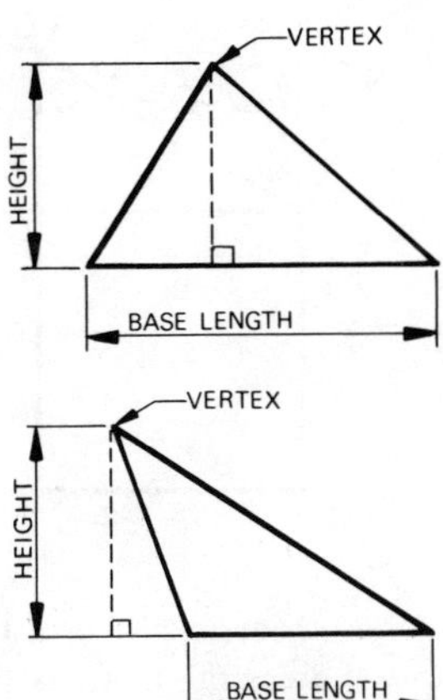

Fig. 2.16 Dimensions of triangles.

Fig. 2.11 and Fig. 2.12 illustrate typical applications.

A *Right angled* triangle has one angle of 90°. The side positioned opposite to the right angle is known as the Hypotenuse and this is the longest side. Fig. 2.13 illustrates a right angled triangle.

An *Isosceles* triangle has two equal angles and two sides of equal length as shown in Fig. 2.14.

Fig. 2.15 shows an *Equilateral* triangle with equal angles of 60°. All sides in an Equilateral triangle are equal in length.

In Fig. 2.16, the method of dimensioning the vertex height from a given base is indicated.

Pythagoras' theorem

This theorem states that in any right angled triangle, the square on the hypotenuse is equal to the sum of the squares on the other two sides.

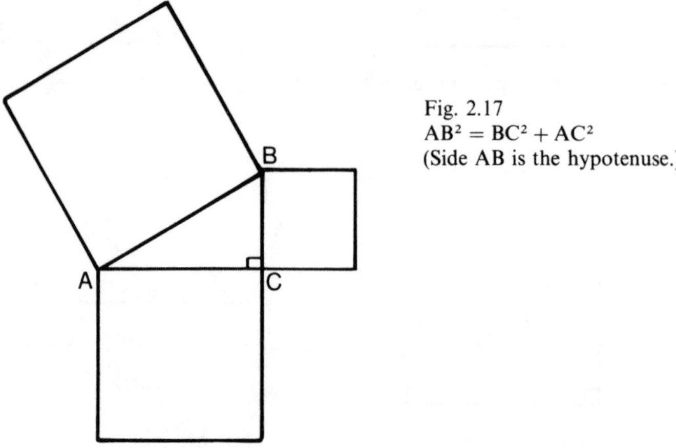

Fig. 2.17
$AB^2 = BC^2 + AC^2$
(Side AB is the hypotenuse.)

Pythagoras' theorem has several interesting applications. It can be extended to the areas of circles as shown in Fig. 2.18 where Circle area A = Circle area B + Circle area C.

The area of the square ABCD in Fig. 2.19, and shown shaded, is half of the area of the square ACEF.

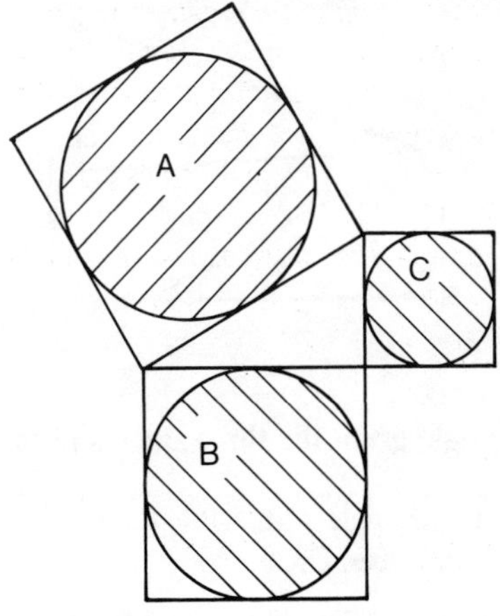

Fig. 2.18

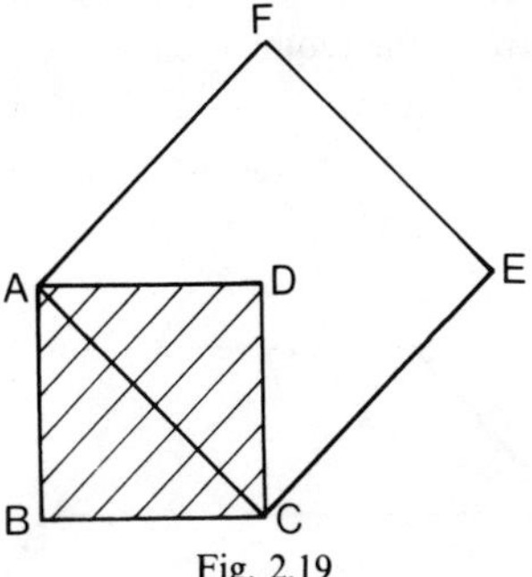

Fig. 2.19

A mathematical application is given in Fig. 2.20 where side AB of the triangle is drawn to represent 1 unit, the base of the triangle represents 2 units and the hypotenuse represents the square root of 5 units. Now if the hypotenuse is the base of a new right angled triangle ACD and the vertical CD is equal to 1 unit, then the new hypotenuse of triangle ACD represents the square root of 6 units. Similarly in triangle ADE, the hypotenuse AE represents the square root of 7 units. Measure the lengths of the sides AC, AD and AE to give the required values. As a further exercise, find the square root of 3 and 4 by commencing the construction with sides AB and BC equal to 1 unit.

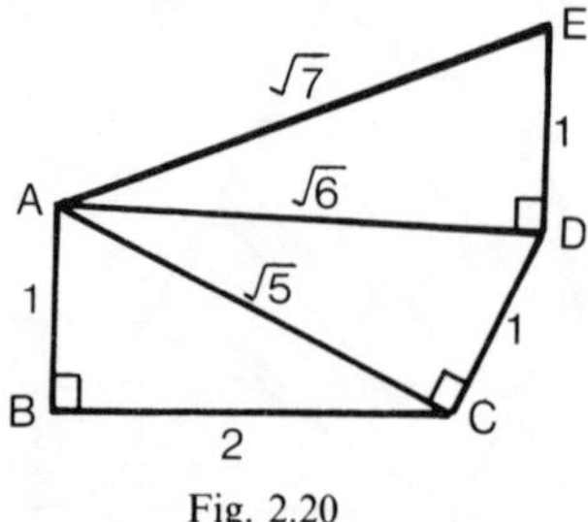

Fig. 2.20

To construct a triangle given the three sides and to draw the inscribed circle

Example Construct a triangle with AB = 80 mm, AC = 70 mm and BC = 60 mm. Draw the inscribed circle.

Solution Fig. 2.21. Draw AB = 80 mm and from point A draw an arc of 70 mm to intersect with an arc of 60 mm from point B. The intersection will give point C. Bisect angles CAB and ABC and the intersection will give point O, the centre of the required circle.

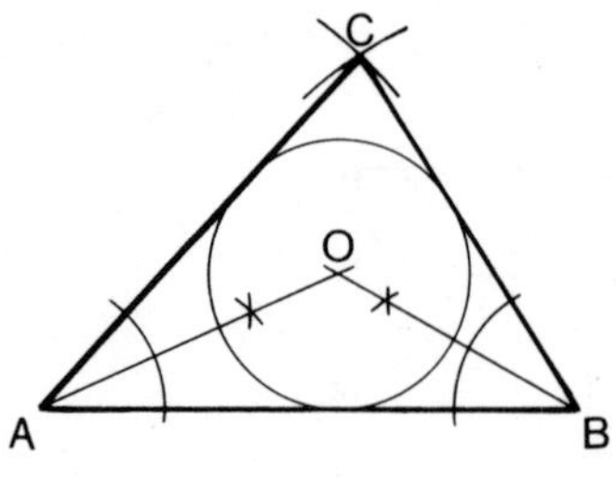

Fig. 2.21

To construct a triangle given one side and two angles and to draw the circumscribing circle

Example Construct triangle ABC with AB = 75 mm, angle CAB = 45° and angle CBA = 60°. Add the circumscribing circle.

Solution Fig. 2.22. Draw AB = 75 mm. Draw an angle of 45° at point A and an angle of 60° at B. The intersections of the sides will give the apex C. Now the centre of a circumscribing circle must be equidistant from all three corners so bisect any two sides and the intersection of the bisecting lines will give the required circle centre, point O.

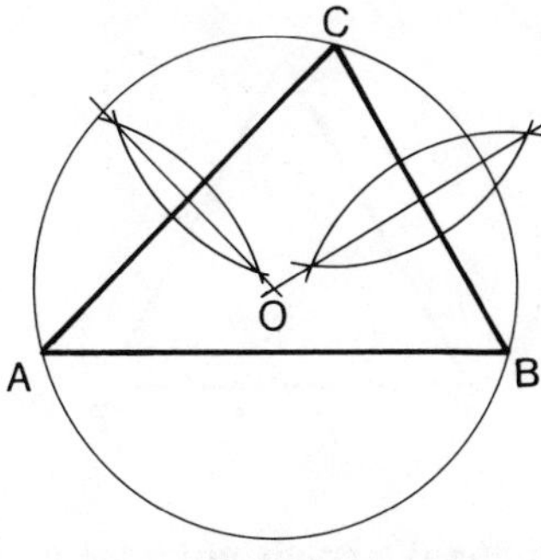

Fig. 2.22

To construct a right angled triangle if the hypotenuse and one other side is given

Example Draw a right angled triangle with hypotenuse AB = 85 mm and side AC = 45 mm.

Solution Fig. 2.23. Draw AB = 85 mm and bisect this length. Draw a semicircle with AB as diameter. From A describe an arc of radius 45 mm to give point C. Join AC and CB to give the required triangle. The angle ACB in the semicircle will be a right angle.

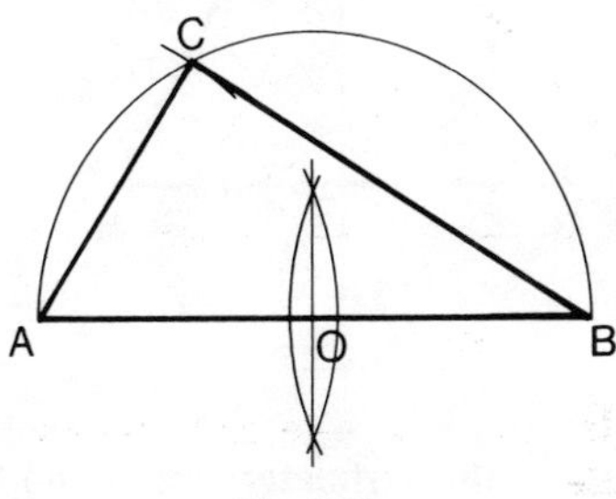

Fig. 2.23

To construct an equilateral triangle given the length of the side

Example Draw an equilateral triangle with a side of 60 mm.

Solution Fig. 2.24. Draw AB = 60 mm. With centre A and radius AB, draw an arc to intersect with a similar arc drawn from point B. The intersection gives point C. Join AC and BC to give the required triangle.

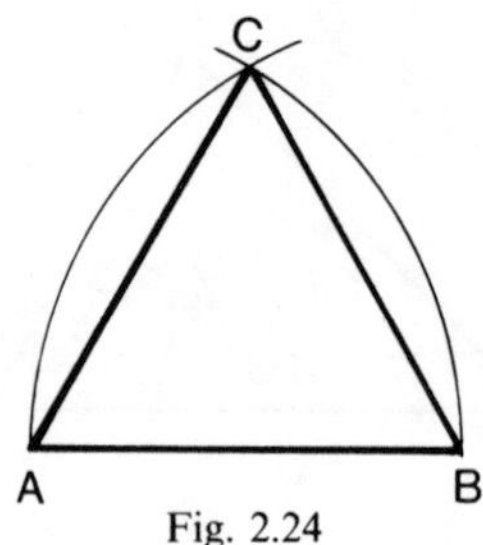

Fig. 2.24

To construct an equilateral triangle given the altitude

Example Construct an equilateral triangle ABC where the altitude AD = 60 mm.

Solution Fig. 2.25. Draw AD = 60 mm. On either side of AD, construct angles of 30°. Construct a perpendicular at point D and extend the line to give intersections B and C.

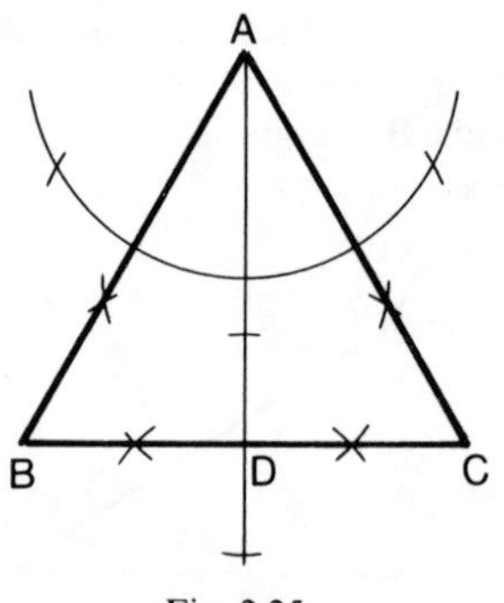

Fig. 2.25

To draw a triangle given the perimeter length and two base angles

Example Draw triangle ABC where the perimeter length is 150 mm, angle ABC = 60° and angle BCA = 45°.

Solution Fig. 2.26. Draw the perimeter length DE = 150 mm. At point D, construct an angle of 30° which is half of one of the required base angles and at E, construct an angle of $22\frac{1}{2}$°, which is half of the other base angle. Let the two sides of the constructed angles intersect at point A. At point A, copy these two angles as shown to form the required triangle.

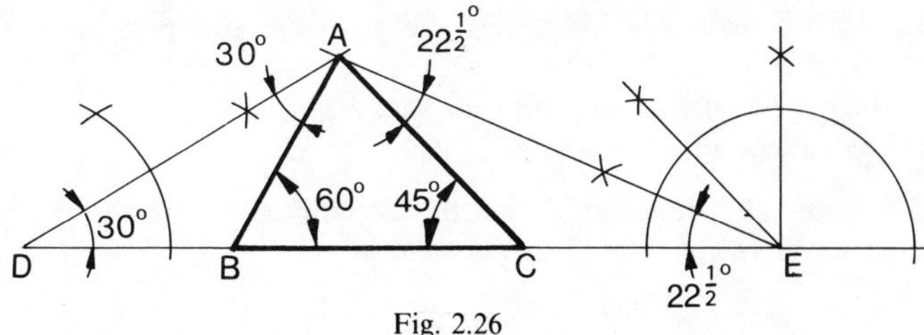

Fig. 2.26

Note that this construction is based on the properties of the two isosceles triangles which are formed, ABD and AEC.

Parallelograms

A *parallelogram* is a rectilinear (i.e. straight lined) four sided figure, where opposite sides are equal in length and also parallel with each other.

The geometrical properties of any parallelogram which are useful to remember are as follows:

(*a*) Diagonals bisect each other.

(*b*) Either diagonal will divide the figure into two similar triangles of equal area.

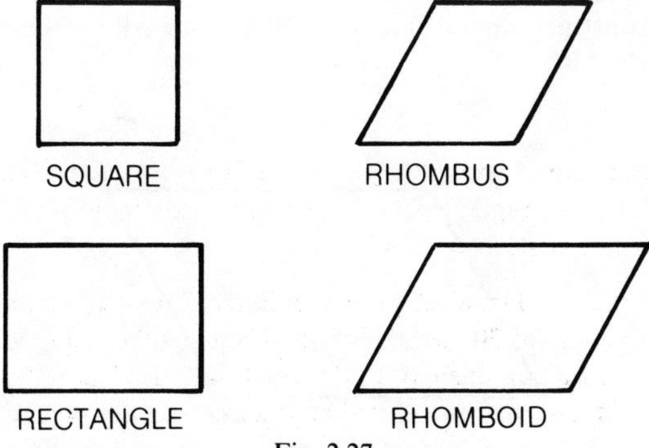

Fig. 2.27

(*c*) The two diagonals will divide the area into two pairs of similar triangles.
(*d*) Opposite angles of any parallelogram are equal in size.
(*e*) The sum of adjacent angles is 180°.

Fig. 2.27 are all parallelograms but this general term is mainly used with reference to the *rhombus* and the *rhomboid.*

The quadrilateral

The name *quadrilateral* describes any figure with four straight lines as its boundary, but where two of the sides are known to be parallel, then the name *trapezium* is invariably used. From the properties previously stated, it will be obvious that a quadrilateral will not necessarily be a parallelogram.

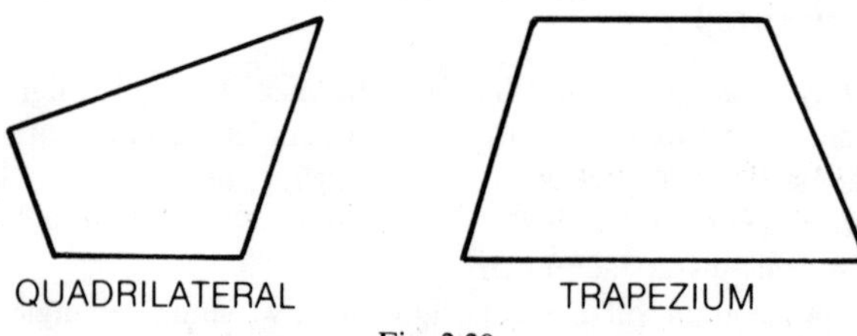

Fig. 2.28

Construction for a parallelogram

Example Construct a parallelogram ABCD with AB = 50 mm, BC = 70 mm and angle ABC = 60°.

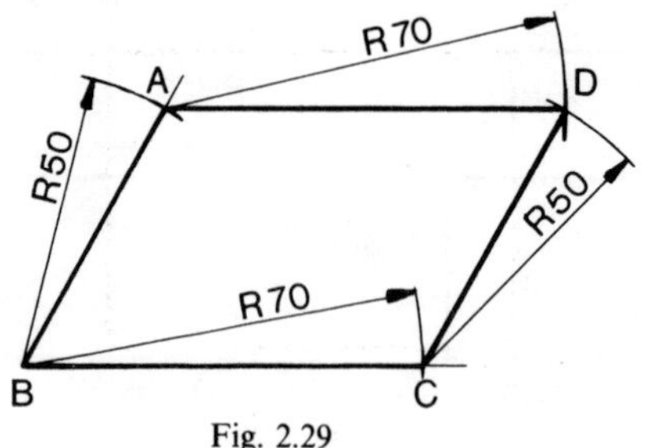

Fig. 2.29

Solution Draw an angle of 60° at B. With centre B and radius of 50 mm, position point A. Similarly, with radius of 70 mm, establish point C. From A, describe an arc of 70 mm to intersect with an arc of 50 mm from C. The intersection gives point D.

Note, to construct the angle of 60° without the aid of a set-square or protractor, draw an arc from point B, Fig. 2.30, and with the same radius, step off an arc from the horizontal line as shown. Draw a line from B through the intersection. This construction for an angle of 60° is based on the fact that a radius can be stepped off six times around the circumference of a circle, giving six equal angles of 60°.

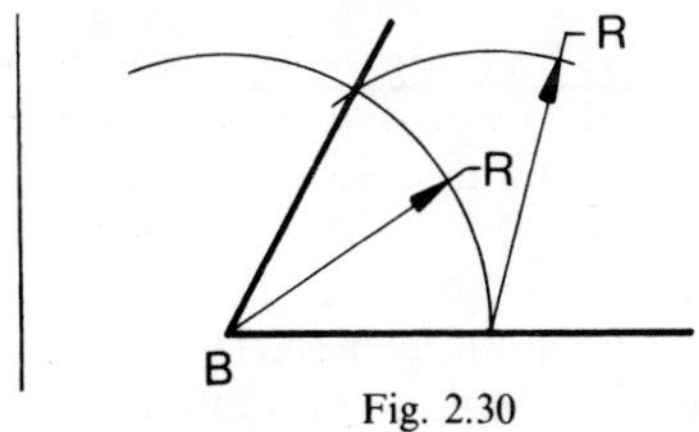

Fig. 2.30

Construction for a rectangle

Example Construct a rectangle ABCD where AB = 50 mm and BC = 70 mm.

Solution Fig. 2.31 shows the construction for an angle of 60° carried one stage further to form a right angle at A. Position points B and D which are 50 mm and 70 mm from A. Point C is positioned by drawing intersecting arcs of 50 mm and 70 mm from D and B.

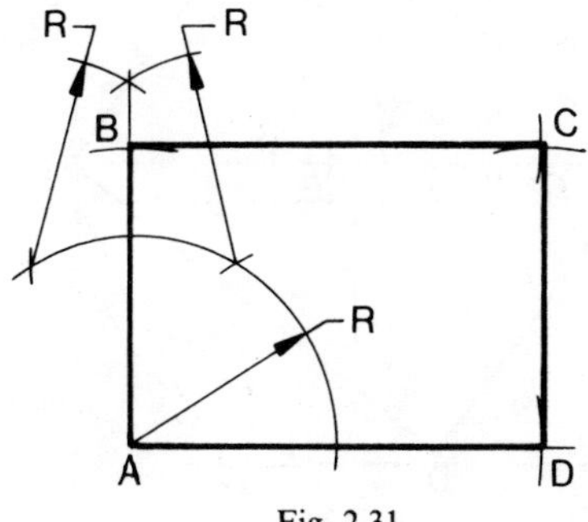

Fig. 2.31

An alternative construction is shown in Fig. 2.32. This method uses the geometrical fact that an angle drawn in a semicircle is a right angle, where the hypotenuse is the diameter.

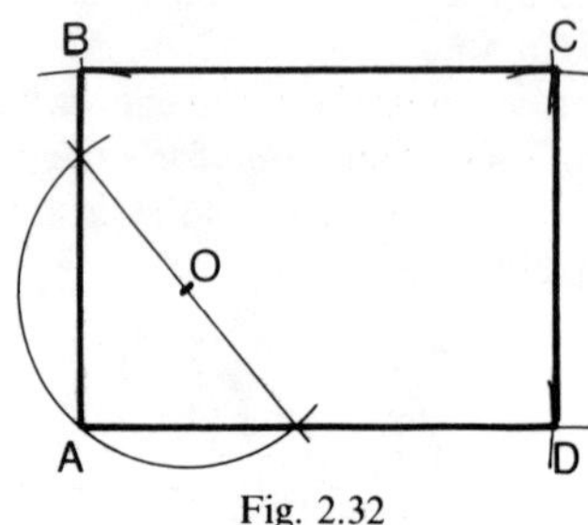

Fig. 2.32

From any point O, draw an arc of convenient radius. Point A lies on the circumference and if lines are drawn from A through the intersection of any diameter with the circle arc, then a right angle will be formed at A.

Position points B, C and D by intersecting arcs as before.

Example Construct a rectangle ABCD, where diagonal AC = 100 mm and side AB = 80 mm.

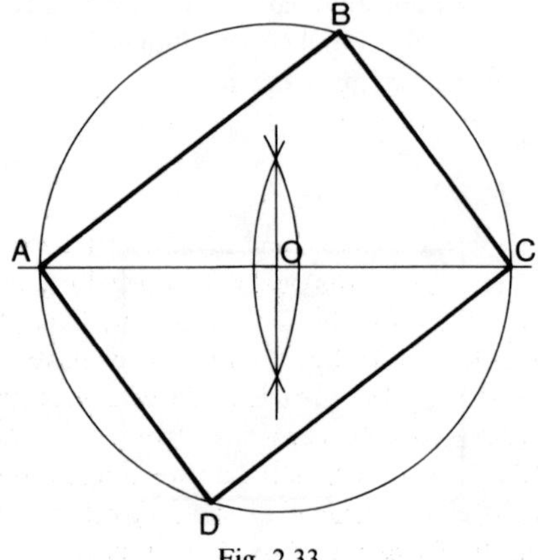

Fig. 2.33

Solution Draw AC and bisect to give point O. Draw a circle with radius OA. From A, draw a radius of 80 mm intersecting the circumference to give point B. Repeat from point C to give point D.

Polygons

An irregular polygon is a figure bounded by straight lines of different lengths. A regular polygon is a figure which is bounded by straight lines of equal length and the angles between them are also equal.

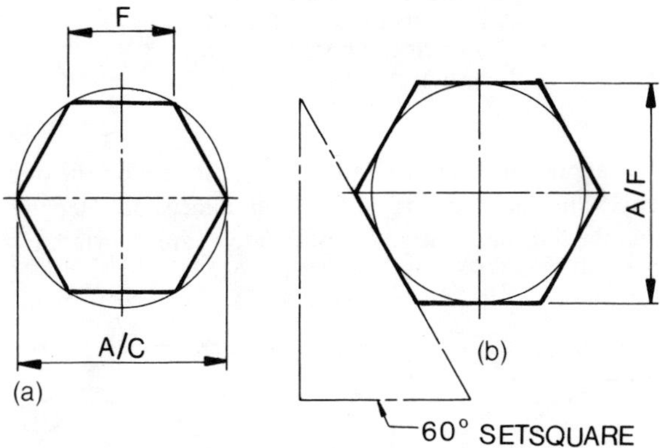

Fig. 2.34

(*a*) Construction given distance across corners.
(*b*) Construction given distance across flats.

A typical regular polygon in common use in technical drawing is the regular hexagon. The size of the hexagon is usually given by the distance across the corners (A/C), the distance across the flats (A/F) or the width of one of the flat sides (F). In Fig. 2.34 hexagons have been drawn with a 60° set square on the inside or the outside of circles, the circle diameter being the dimension across the corners or flats.

In the case of the hexagon where the distance across the corners is given, the hexagon can also be constructed by stepping off the circle radius around the circumference. It will be found to go six times.

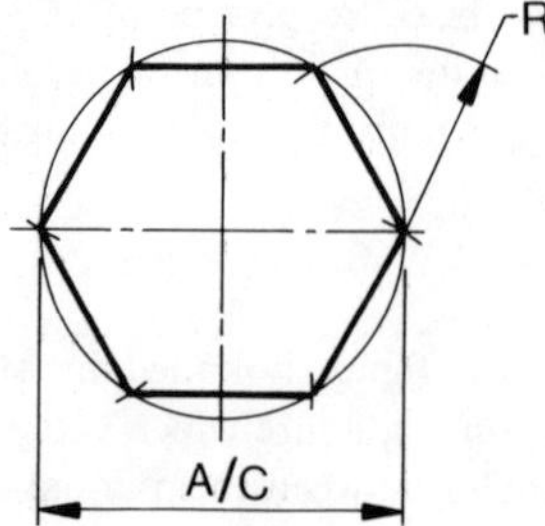

Fig. 2.35 Alternative construction given distance across corners.
A/C is distance across corners
A/F is distance across flats
F is width of flat side
R is circle radius

A regular octagon can be drawn with a 45° set square on the outside or inside of a circle and this depends on the given dimension referring to the distance across the flats or corners. The alternatives are shown in Fig. 2.36.

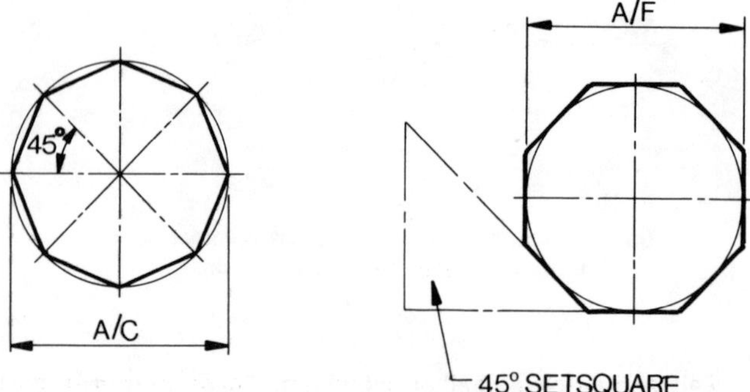

Fig. 2.36 Construction of a regular octagon.
A/C is distance across corners
A/F is distance across corners

Fig. 2.37 shows a method of drawing a regular polygon, given the size of the enclosing circle. For example, draw a regular polygon with seven sides inside a circle of 80 mm diameter.

Draw diameter AB, 80 mm long and the circle. Divide AB into the same number of parts as the sides in the regular polygon. With centre A, draw an arc equal to the diameter to intersect with another arc of the same radius, but drawn from centre B. The intersection gives point C. Draw a line from point C through the second division along the diameter, to intersect the circumference at point P. The line AP is one side of the required polygon. Step off arcs equal in length to AP around the circumference to give the other sides. Note that in this construction, the line from point C always passes through the second division along AB, irrespective of the number of sides in the required polygon.

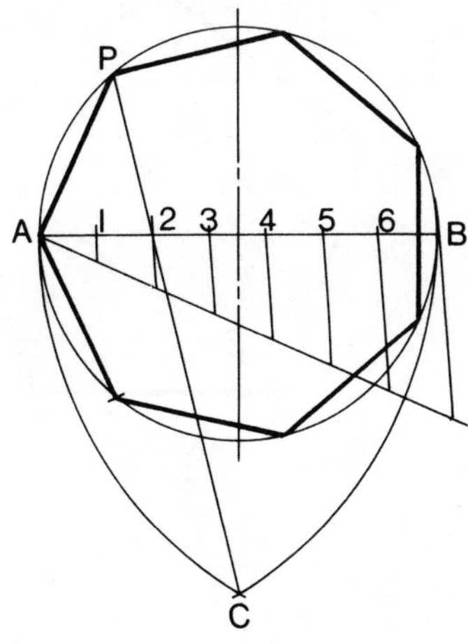

Fig. 2.37

Note: The method of dividing a line into an equal number of parts is given later in this chapter in the section on scale drawing. Please refer to Figs 2.56 and 2.57 and their explanations as the principle is used in several problems which follow.

To draw a regular polygon, given the length of one side

Example Draw regular polygons with five and seven sides, where the length of the side in both cases is 50 mm.

Solution Fig. 2.38 shows the five sided polygon, known as a *pentagon.*

Draw the known side AB, 50 mm in length, and using this length as a radius, describe a semicircle. Divide this semicircle into the same number of equal sectors as sides in the required polygon. Line A2 is the second side in the polygon. Draw line A3 and extend for a convenient distance. With radius AB and centre on point 2, describe an arc to intersect with the extended line A3, and the intersection will give point D. Similarly extend the line A4 and draw a radius equal to AB using centre D, the intersection gives point C.

Repeat the above procedure for the second polygon and note that irrespective of the number of sides required in the figure, the second side is always

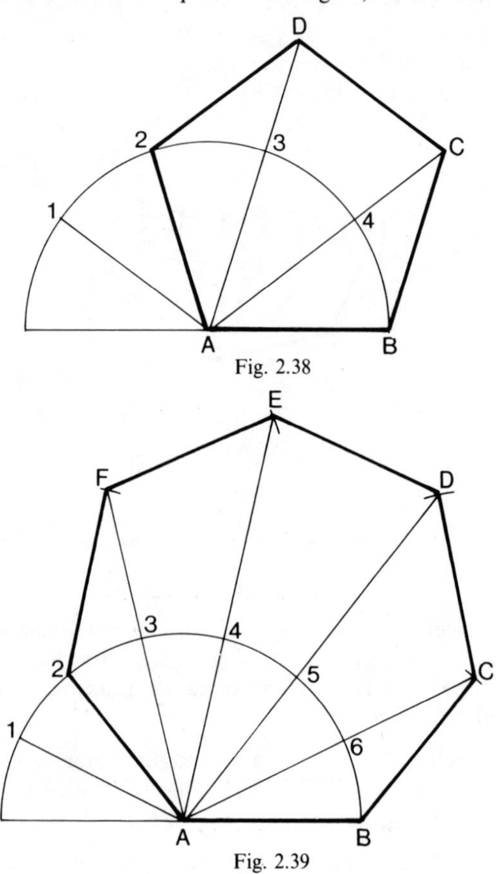

Fig. 2.38

Fig. 2.39

obtained from the centre of the semicircle A to the second division around the circumference. A polygon with seven sides is known as a *heptagon.*

Areas

The amount of space within a boundary is its area and is measured, in the case of a rectangle, by multiplying the length by the height. This measurement is given in square millimetres (mm^2), square centimetres (cm^2), square metres (m^2), or the square of any other linear dimensions which may have been used. One square centimetre is the space enclosed by a square with sides one centimetre in length and is also, for example, the area of a rectangle with 0·5 and 2 centimetre sides.

The area of rectangle ABCD in Fig. 2.40, is its base length AB multiplied by its height CB. Now the diagonal AC bisects the rectangle and hence the area of the triangle ABC will be the base length multiplied by the height divided by two.

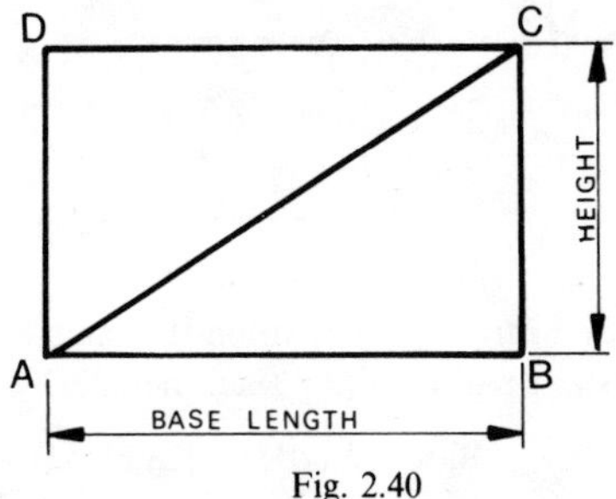

Fig. 2.40

In the case of the parallelogram ABCD, in Fig. 2.41, it can be seen that

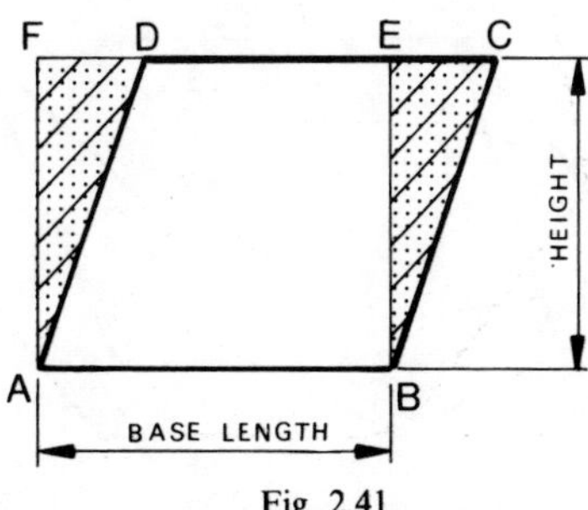

Fig. 2.41

the areas of the two shaded triangles are identical since the height is common to both and FD = EC, as AD and BC are parallel. It follows that the area of the parallelogram and rectangle are the same, namely, base length multiplied by perpendicular height.

It also follows that parallelograms which are constructed on the same bases or drawn with equal bases are equal in area provided they are contained between equidistant parallel lines.

$$\text{Parallelogram area} = \text{Base} \times \text{Height } (\text{B} \times \text{H})$$

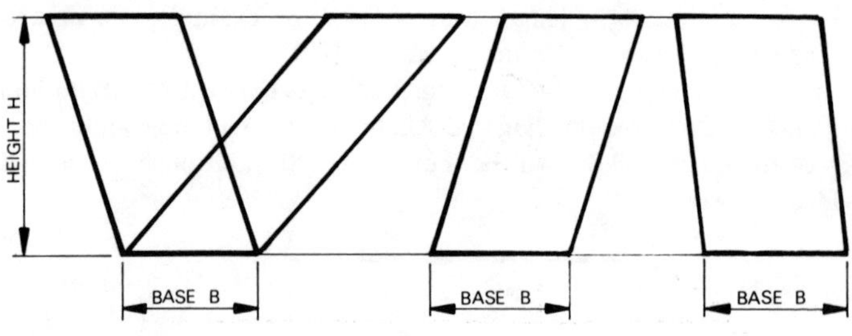

Fig. 2.42

Similarly, in the case of triangles drawn on the same base, or with equal bases between equidistant parallel lines, then the areas will be equal.

$$\text{Triangle area} = \frac{\text{Base} \times \text{Vertical Height}}{2} \left(\frac{\text{B} \times \text{H}}{2}\right)$$

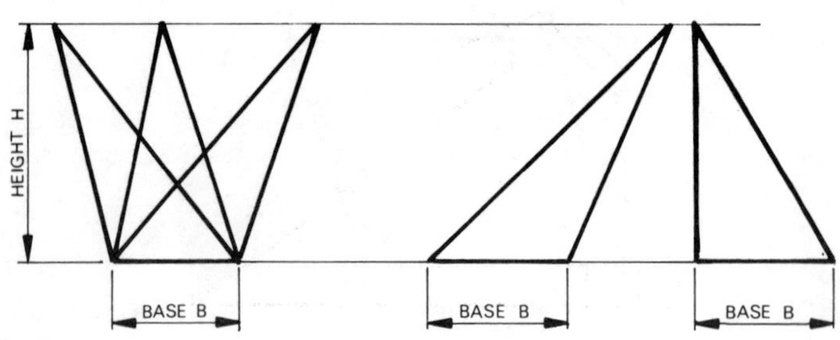

Fig. 2.43

Areas of polygons

The areas of multi-sided figures can be determined by reducing them to triangles of similar area by the following construction.

Example Find the area of quadrilateral ABCD, where AB = 80 mm, Angle ABC = 60°, BC = 70 mm, CD = 55 mm and AD = 50 mm. First, draw AB, construct angle ABC and measure length BC. From point C, describe an arc of radius 55 mm to intersect at point D with an arc of radius 50 mm from point A.

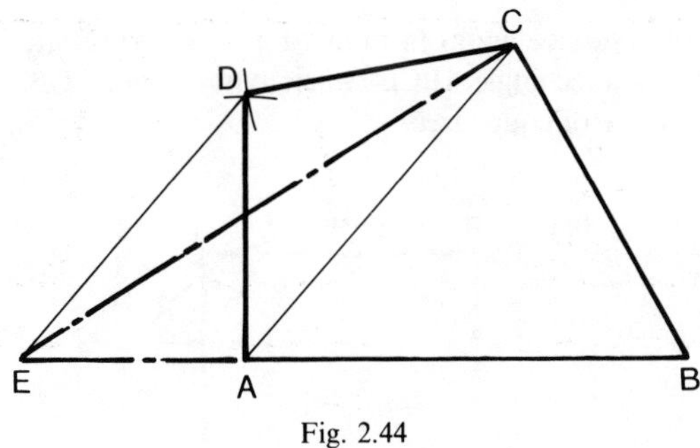

Fig. 2.44

Solution Fig. 2.44. Join CA and draw a line from D parallel to CA to meet BA extended to E. Now triangles ACE and ACD are equal in area, hence the area of the quadrilateral ABCD is equal in area to the triangle BCE

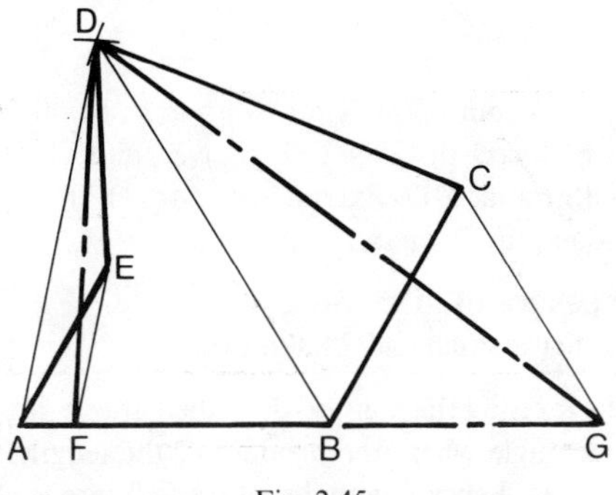

Fig. 2.45

and the number of sides in the original figure has been reduced from four to three.

The area can be found by measuring the length of BE, then measure the vertical distance from BE to the vertex C. Multiply these two dimensions and divide the product by two for the required area.

Fig. 2.45 shows a five sided polygon ABCDE. The construction shown above has been used twice to construct the triangle DFG of similar area. Copy this solution using the following dimensions to draw the original polygon. AB = 60 mm, Angle ABC = 120°, BC = 50 mm, AE = 35 mm, Angle BAE = 60°, ED = 40 mm and CD = 75 mm.

To construct a square equal in area to a given rectangle

Example Draw a rectangle ABCD, where AB = 35 mm, CB = 80 mm and construct a square of equal area.

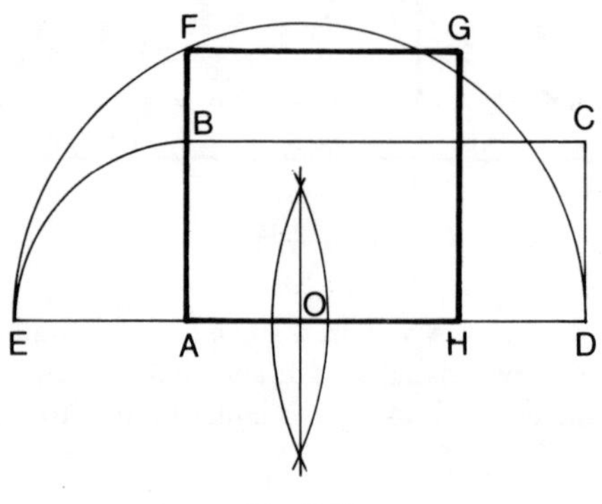

Fig. 2.46

Solution Fig. 2.46. From point A, draw an arc of radius AB to intersect DA extended to E. Bisect distance ED to give point O. Draw a semicircle from centre O with radius OD. Extend AB to touch the circle at F and AF is one side of the required square.

To construct a square of given area

Example Construct a square of 2400 mm².

Solution Fig. 2.47. Using the method described above, it will first be necessary to draw a rectangle where the product of the lengths of the sides will give the required area, hence rectangles of, say 40 mm × 60 mm or 30 mm

× 80 mm will be suitable. In Fig. 2.47 the construction previously described has been drawn with AB = 40 mm and BC = 60 mm.

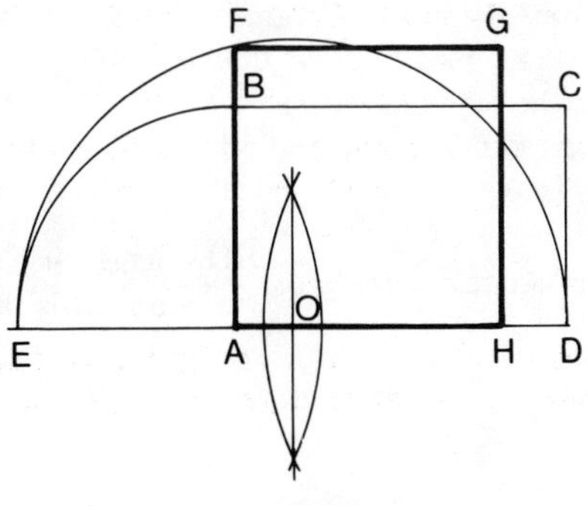

Fig. 2.47

To construct a square equal in area to the sum of several other squares

This construction is an application of the well known theorem of Pythagoras, relating to right angle triangles.

Example Draw a square equal in area to the combined areas of squares having sides of lengths 20, 30, 40 and 60 mm.

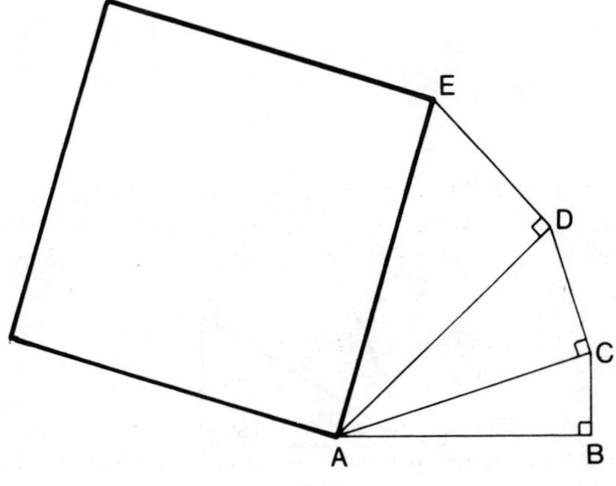

Fig. 2.48

Solution Fig. 2.48. A solution is shown here but it is not the only one since the lengths can be added in any convenient order. Draw AB = 60 mm, CB = 20 mm and perpendicular to AB. The square on AC represents the sum of the squares on AB and BC. Continue by drawing DC perpendicular to AC and 30 mm long. Finally draw ED perpendicular to AD and the square on AE gives the required area.

The method described in the above solution can also be used to find the sums of circular areas.

The area of a circle is ΠR^2, or $\frac{\Pi D^2}{4}$ where R is the radius and D is the diameter.

Note that $R = \frac{D}{2}$ and therefore $\Pi R^2 = \Pi \left(\frac{D}{2}\right)^2 = \frac{\Pi D^2}{4}$.

$\frac{\Pi D^2}{4}$ is the expression generally used by technicians and engineers. Try measuring the radius of a hole to appreciate why it is simpler to work with the distance across the diameter.

Example Find the total cross sectional area of circular pipes having diameters of 30 mm, 40 mm and 50 mm.

Solution Draw right angled triangles in any convenient order, for example, as shown in Fig. 2.49 and the required area for the total cross section will be $\frac{\Pi}{4}(AD)^2$.

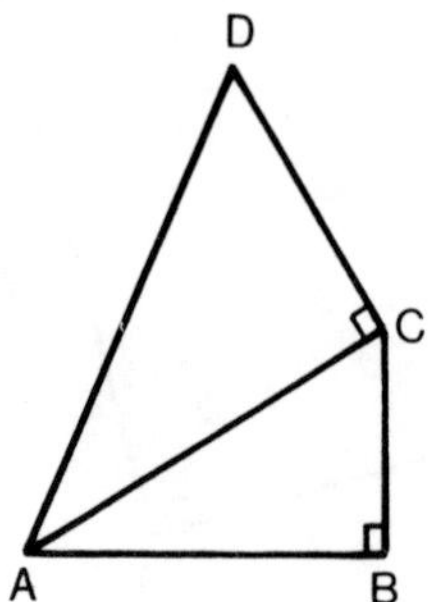

Fig. 2.49 AB = 50 mm, BC = 30 mm and DC = 40 mm.

The Greek letter Π (which is pi) is used as a symbol of the ratio of circle circumference to diameter. The approximate value of Π is usually taken as 3·14 or $\frac{22}{7}$.

It is often necessary to reproduce areas to a scale of twice full size or half full size and the Pythagoras principle can be used to solve this type of problem.

Example Draw a quadrilateral ABCD where AB = 50 mm, BC = 60 mm, CD = 70 mm, AD = 85 mm and Angle ABC = 120°. Draw similar quadrilaterals which are (*a*), twice the original area in size and (*b*), half the original area in size.

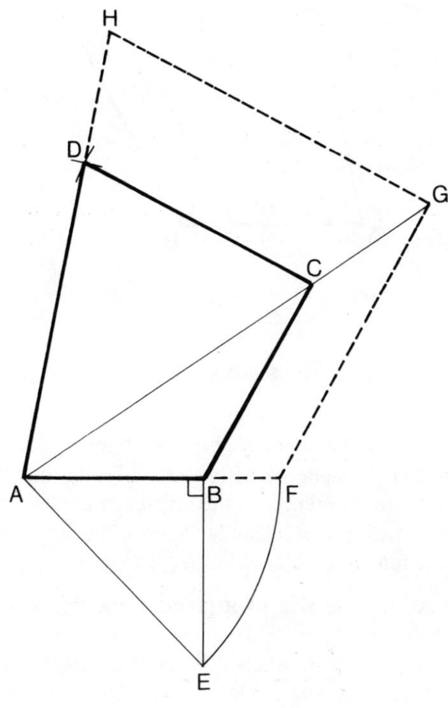

Fig. 2.50

Solution (*a*) Draw BE at 90° to AB and equal in length to AB. From centre A, draw radius AE to intersect AB extended to F. Draw FG parallel to BC and GH parallel to CD. The required area is AFGH. Check the area of your solution by calculating the sums of the separate component triangles AFG and AGH. Remember that the triangle area is equal to its base length multiplied by perpendicular height divided by two.

Solution (*b*) Draw the perpendicular bisector of AB at point O. OA = OE. Draw radius AE to give point F. Draw FG parallel to BC and GH parallel to CD. AFGH is the required area. Check its value as before.

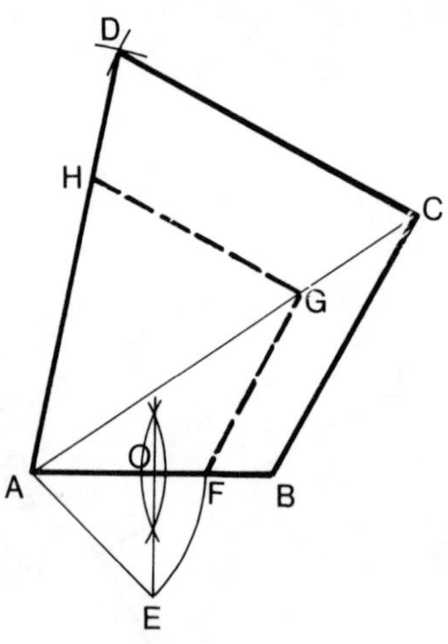

Fig. 2.51

It will be appreciated that similar figures are figures which have the same shape but are not of the same size. Problems relating to enlargements and reductions can refer to differences in the lengths of the lines which form the figure, or to differences in area of similar figures. This is an important point since the constructions in each case are not the same.

To enlarge or reduce the size of a given plane figure given the ratio of the sides.

Example Draw figure ABCD where AB = 40 mm, BC = 20 mm, AD = 35 mm, Angle DAB 60° and Angle ABC = 135°. Draw similar figures in the ratio 1:3 also 4·5:3.

Solution Fig. 2.52. Set out the figure ABCD from the dimensions given. This solution shows a radial or polar method of solving the problem which is based on similar triangles. Let P be any convenient point and draw the line PA. Divide this line into three parts. Draw lines PD, PC and PB. Insert line A_2D_2 which is parallel to AD. Now PA_2 and PA are in the ratio 1:3 and hence A_2D_2 is one-third the length of AD. Draw the other three sides of the reduced figure parallel to those in the given figure. For the enlargement, extend the radial lines from the point P through A, B, C and D. Extend the scale to 4·5 and establish point A_1. Draw A_1D_1 parallel to AD and note that the line is one and a half times greater than AD. Draw the other three sides parallel to DC, BC and AB, to give the required enlargement.

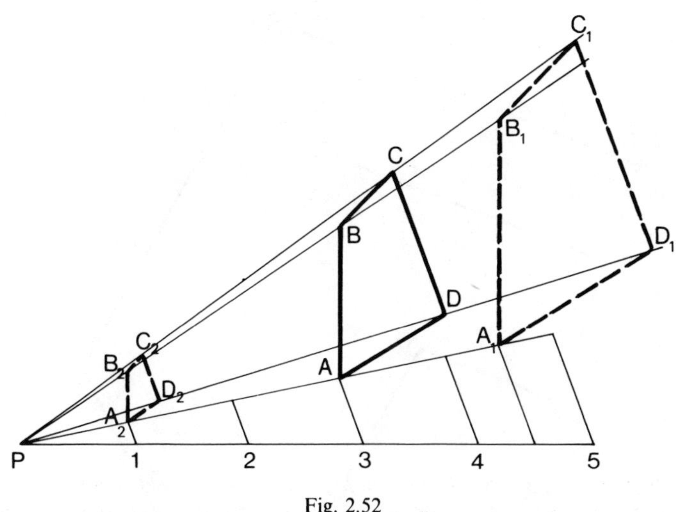

Fig. 2.52

An alternative method can be used where the pole is positioned at one of the corners of the figure.

Example Draw polygon ABCDE, where AB = 70 mm, BC = 40 mm, CD = 80 mm, Angle ABC = 120°, Angle BCD = 120°, DE = 80 mm and AE = 80 mm. Enlarge the polygon so that the ratio of the sides is 4:3 and reduce the polygon so that the ratio is 1:2.

Solution Set out the polygon as shown in Fig. 2.53 and extend the side AB for convenient distance. Draw a vertical line from point B. Place the end of a scale rule at point A and adjust the position of the rule so that six equal divisions span the distance between point A and the vertical line from B. Draw a line against the rule and mark the six divisions. Extend this line to give points 7 and 8. The vertical line from point 8 meets the extension of line AB at B_1. A vertical line from point 3 gives B_2. The lines B_1C_1 and B_2C_2 are drawn parallel to the given line BC and this procedure is continued to complete the new polygons.

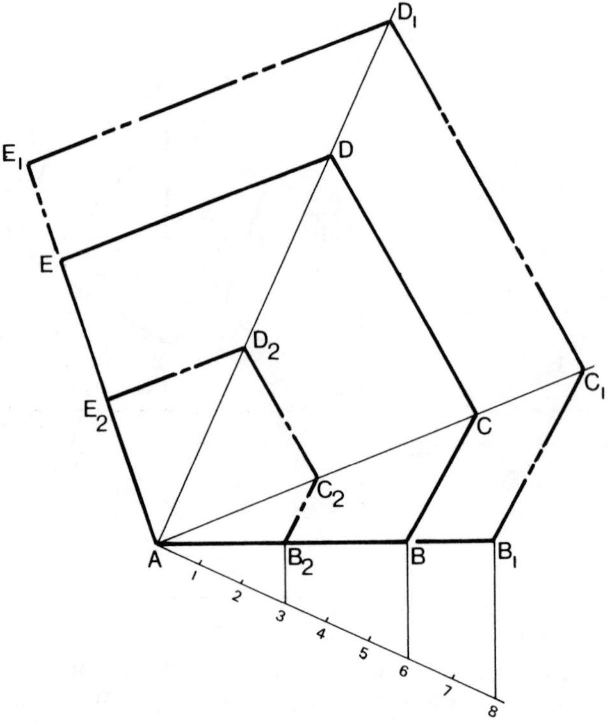

Fig. 2.53

To enlarge the area of a given plane figure by a given ratio

Example Draw polygon ABCD where AB = 60 mm, angle ABC = 110°, BC = 40 mm, Angle BCD = 115° and CD = 40 mm. Enlarge the figure in the ratio 8:5.

Solution Set out the polygon as shown in Fig. 2.54. Extend the line BA a convenient distance and draw the semicircle with AB as radius to give point E. Now distances AF and AE are required to be in the ratio of the enlargement, hence the scale construction which is shown from point A. Bisect line BF to give point O and draw a semicircle of radius OB. A perpendicular line drawn from A will intersect the two semicircles at G and H. Draw HB_1 parallel to GB giving the first extended side of the enlarged figure which is AB_1. Add B_1C_1 parallel to BC also C_1D_1 parallel to CD.

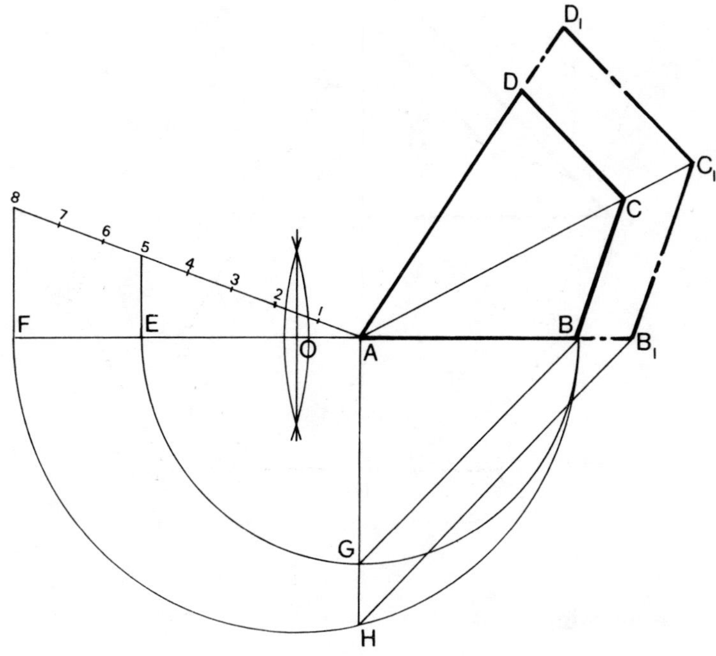

Fig. 2.54

To reduce the area of a plane figure by a given ratio

Example Draw a right angled triangle ABC where CA = AB = 70 mm and Angle CAB = 90°. Draw a similar triangle reduced to three-fifths of the original area.

Solution Set out the triangle as shown in Fig. 2.55 and draw a semicircle with AC as radius, centre A. Divide AD into five equal parts and position E so that AE is three-fifths of AD. Bisect distance CE to give point O and draw a semicircle with O as centre and radius OC. Drop a perpendicular line from A to intersect the two semicircles at F and G. Draw C_1F parallel to CG and C_1B_1 parallel to CB. AB_1C_1 is the required triangle reduced in area to three-fifths of the original area of triangle ABC.

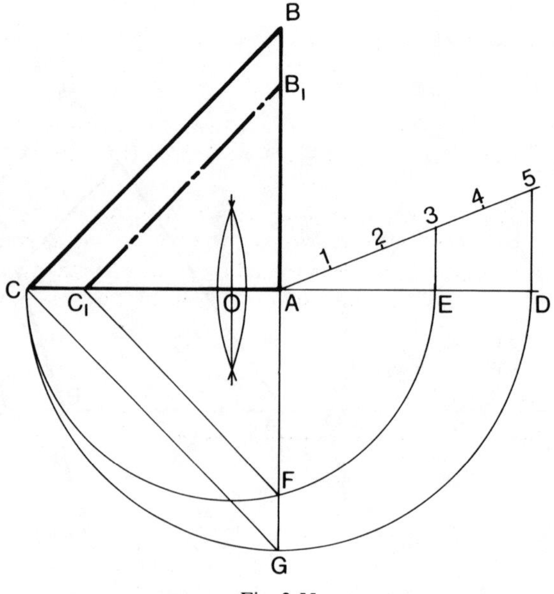

Fig. 2.55

Scale drawing

Generally, it is easier to produce and understand drawings if they represent the true size of the object. This is of course not always possible and it is often necessary to draw enlargements of very small objects and reductions of very large ones. Drawings are also printed by methods which can enlarge or reduce the actual reproductions, so knowledge of the true size of the subject matter on the drawings is most desirable.

All drawings should be drawn in proportion, that is to a uniform scale, and the scale used should be stated on the drawing as a ratio. Normally, scale multipliers and divisors of 2, 5 and 10 are recommended. Scales in common use are as follows:

1000:1	50:1	1:1	1:50
500:1	20:1	1:2	1:100
200:1	10:1	1:5	1:200
100:1	5:1	1:10	1:500
	2:1	1:20	1:1000

The scale on the drawing should be quoted in this manner and so if the drawing has been drawn at five times full size, the scale will be indicated as 5:1.

Scale rules are available, suitably calibrated, for drawing enlargements and reductions but these are special items of equipment and it may be necessary to construct a particular scale. There are two types of scale: plain and diagonal scales. Both of these scales can be used for enlargements and reductions. The plain scale is the simplest to draw but the diagonal scale is used for more accurate scale lengths and examples of both types follow.

To construct a scale, it is necessary to be able to divide a given distance

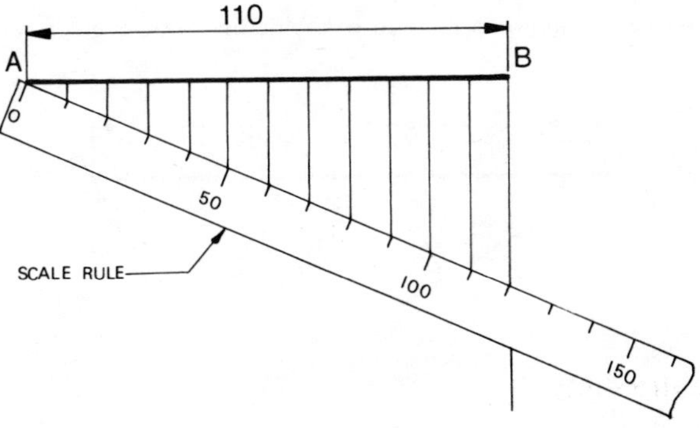

Fig. 2.56

into a number of parts and if the parts cannot be taken readily from a scale rule, then one of the following methods can be used.

To divide a line into a given number of parts

Example Divide a line 110 mm long into twelve equal parts.

Solution Fig. 2.56. Draw the line AB = 110 mm. Add a perpendicular line from B. Place a scale rule on the drawing so that twelve equal divisions span the distance between A and the vertical line from B. Draw a line against the rule and mark the divisions on it. Draw vertical lines from each of these points to intersect the given line AB.

An alternative method is shown in Fig. 2.57 where a line of 100 mm is divided into seven equal parts. Draw line AB = 100 and add the line AC at any convenient angle, but say approximately 30°. Mark along AC seven equal divisions and number them. Join point 7 to point B and with the aid of two set squares, draw parallel lines through points 1 to 6 to line B7.

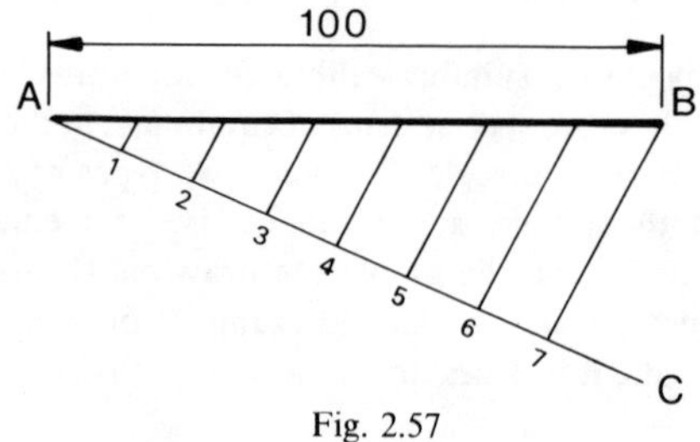

Fig. 2.57

The above methods can be used to divide lines into a given ratio.

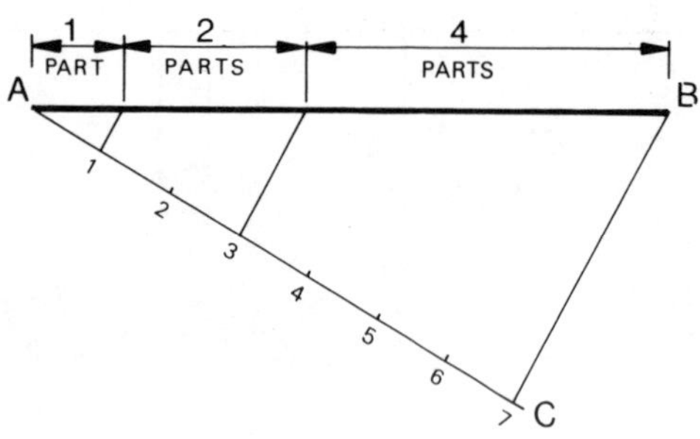

Fig. 2.58

Example Divide a line AB, 120 mm long into the ratio 1:2:4.

Solution Fig. 2.58. The numbers in the ratio add up to seven, so draw AB and divide line AC into seven parts. Draw lines parallel to B7 through points 1 and 3 to intersect the given line.

A further extension can be used to draw a triangle with sides in a given ratio, if the length of the perimeter is known.

Example Draw a triangle ABC, where the three sides are in the ratio of 2:3:4 and the perimeter length is 170 mm.

Solution Draw a line of 170 mm and divide it into nine equal parts as shown in Fig. 2.59. Let BC be a side of the triangle. Draw intersecting radii of the two parts and four parts from opposite ends of line BC to give point A. Join AB and AC.

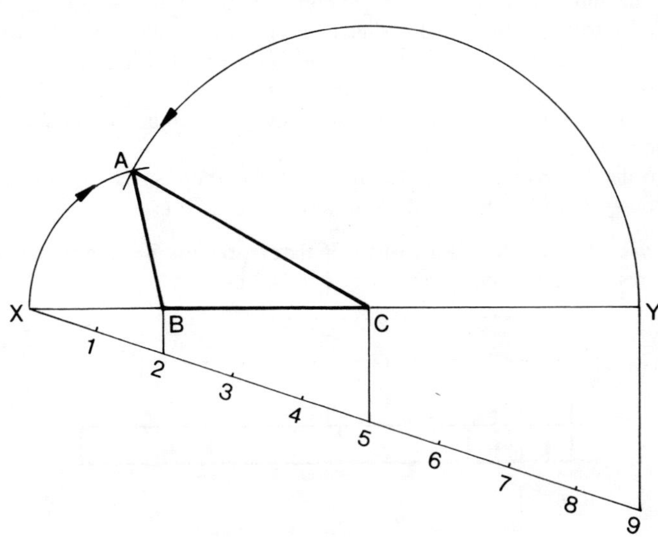

Fig. 2.59 XY = 170 mm.

Plain scales

The construction of a typical plain scale is shown below. Fig. 2.60 illustrates an example where 30 mm represents 1 metre and the scale can be used to read up to 5 metres in increments of 250 mm.

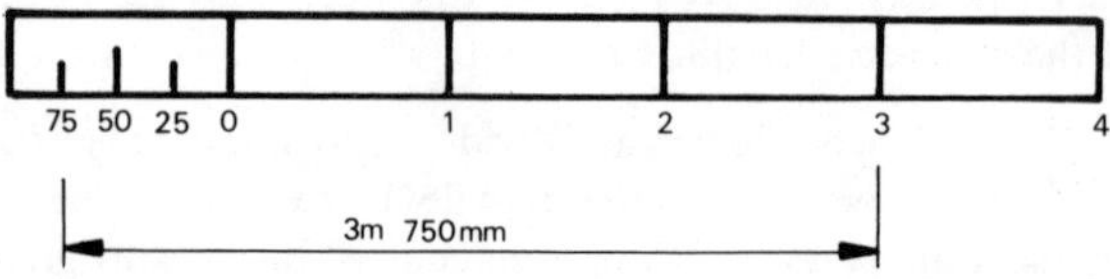

Fig. 2.60 Plain scale where 30 mm represents 1 metre.

Draw a horizontal line 150 mm long and divide this into 30 mm parts. The division on the left hand side is used for the smallest units required in the scale and since it is a unit of 250 mm, then the 30 mm length has been divided into four. Note that the first division is the only one which it is necessary to subdivide and that at the end of the first division 'O' is printed. To take off a scale length, measure from right to left. An example is given for a reading representing 3 m 750 mm. Always remember to quote details of the scale used or the ratio.

In order to subdivide accurately the left hand part of the scale, the methods previously explained for the division of a line can be used.

Example Draw a plain scale where 36 mm represents 500 mm to measure 2 metres. Indicate a measurement of 1 m 200 mm.

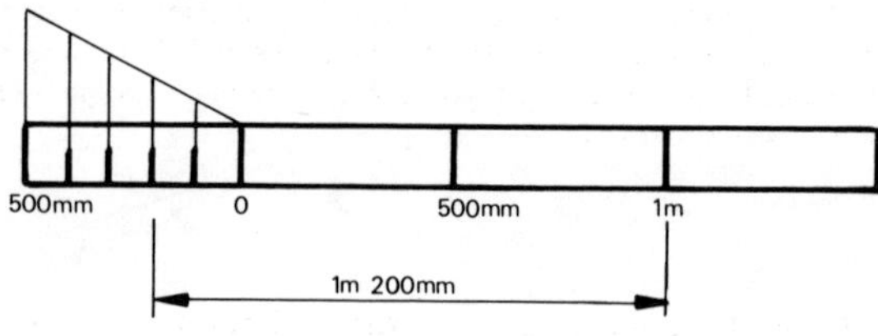

Fig. 2.61

Solution Fig. 2.61. Plain scale where 36 mm represents 500 mm. *Note* that the total length of this scale will be 4 × 36 = 144 mm.

Diagonal scales

Diagonal scales are used to determine more precise scale measurements. The scales take their name from the fact that a series of diagonal lines are used in the sub-division of the smallest part of the scale.

Example Draw a diagonal scale where 50 mm represents 1 metre. It is required to read the scale by 10 mm increments up to a maximum length of 3 metres.

Solution Construct the scale by drawing a line 150 mm long and divide it into three equal parts. Add ten vertical divisions, and these can be to any reasonable dimension. In the construction shown below 5 mm distances were used. Add the diagonal lines indicated in the left hand part of the scale. A typical scale length has been included.

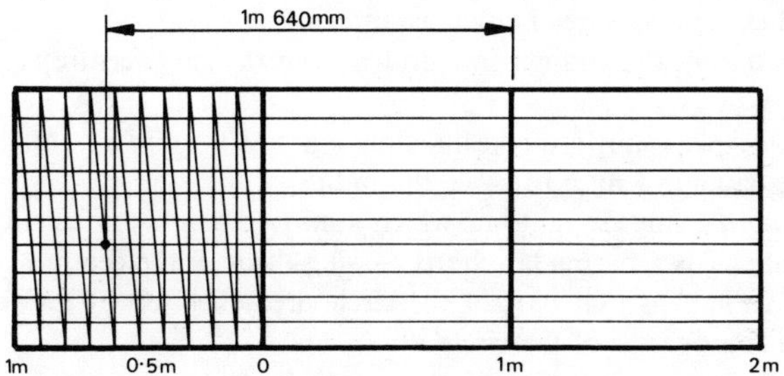

Fig. 2.62 Diagonal scale where 50 mm represents 1 metre. Note that the first metre has been divided horizontally into ten 100 mm units and each has been divided vertically into ten 10 mm increments by the use of the diagonal construction lines.

Example Construct a diagonal scale where 25 mm represents 1 metre and indicates 10 mm, 100 mm and metre units up to 4 m. Show on the solution a reading of 3·55 m.

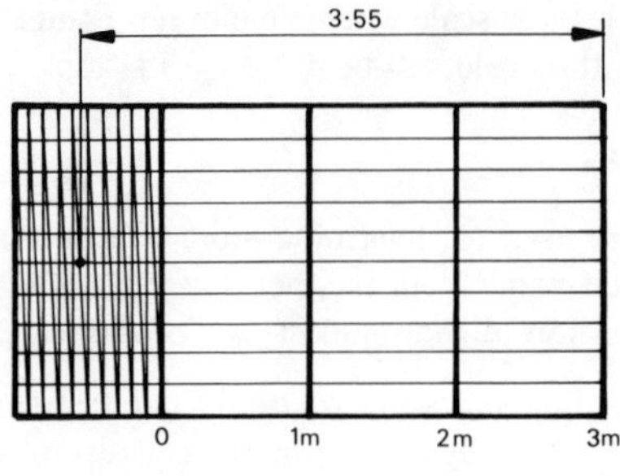

Fig. 2.63

Solution Fig. 2.63. Scale where 25 mm represents 1 metre in 10 mm increments. Note the width of the scale is $4 \times 25 = 100$ mm.

The circle and tangency

A *circle* is a plane figure enclosed by a *circumference* on which every point is equidistant from a fixed point, called the *centre* of the circle. Any part of the circumference is known as an *arc*.

A *radius* is any straight line drawn from the circle centre to the circumference.

A *diameter* is any straight line drawn across the circle through the centre and is equal in length to twice the radius.

A *chord* is any straight line which joins two points on the circumference. A diameter is a particular chord passing through the centre.

A *quadrant* is that part of the circle area enclosed by an angle of 90°, which is a quarter of the circle.

A *semicircle* encloses one half of a circle.

A *sector* is the area bounded by two radii and the arc which they cut.

A *segment* is the area bounded by an arc and a chord.

Concentric circles are circles with different radii, having a common centre.

Eccentric circles are circles which are not concentric. The distance between the circle centres is known as the *eccentricity*.

A *tangent* is a straight line which touches the circumference at a point, known as the *point of tangency*. The tangent does not cut the circle.

A straight line drawn between the circle centre and the point of tangency is known as the *normal*.

The angle between the tangent and the normal is 90°.

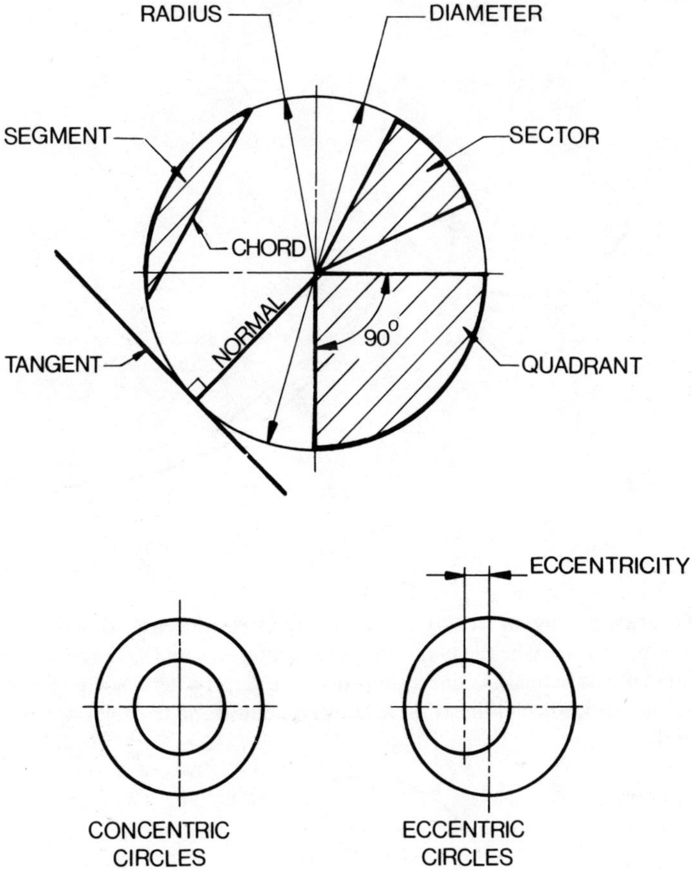

Fig. 2.64

To find the centre of a given circle

The centre of a given circle can be found using the fact that the perpendicular bisector of any chord passes through the circle centre. As shown in Fig. 2.65, draw any two chords and the intersection of the perpendicular bisectors will give the circle centre. The two chords should not of course be parallel to each other. Similarly, the centre of a given arc can be established by adding two chords and bisecting them. In the event that space is limited, it does not matter if the chords overlap, as shown in Fig. 2.66.

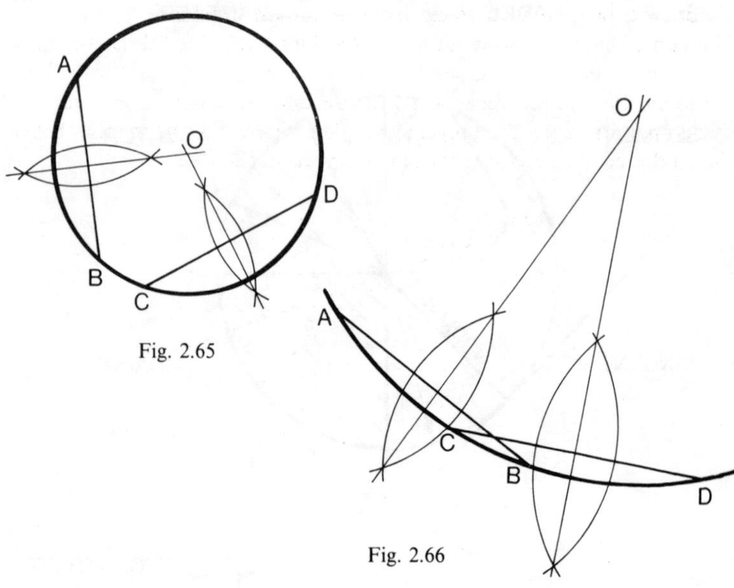

Fig. 2.65

Fig. 2.66

To draw a tangent at any point on the circumference of a circle

In Fig. 2.67, let P be the point on the circumference and O the circle centre. Join OP and extend the line a short distance. Erect a perpendicular at point P. The perpendicular line is the tangent at point P and line OP is the normal to the tangent.

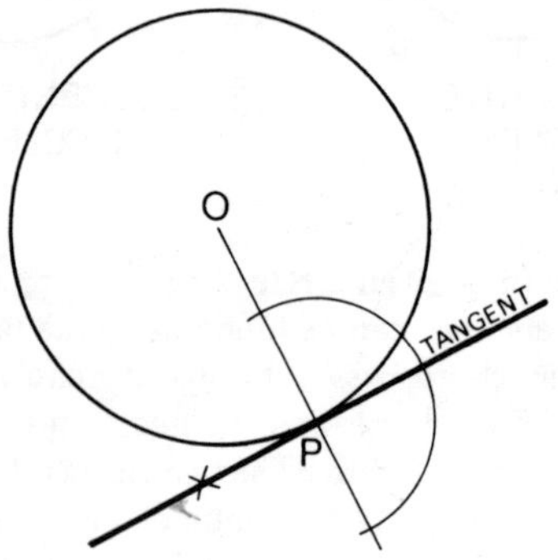

Fig. 2.67

To draw a tangent to a circle from any point outside it

The construction is shown in Fig. 2.68. Draw the line OP to the circle centre and bisect it, giving point Q. Draw a semicircle from Q with radius PQ to cut the circumference at R. Line PR is tangential to the circle and OR is the normal to the tangent. Another intersection is possible below, shown dotted, so two tangents can be drawn in this case.

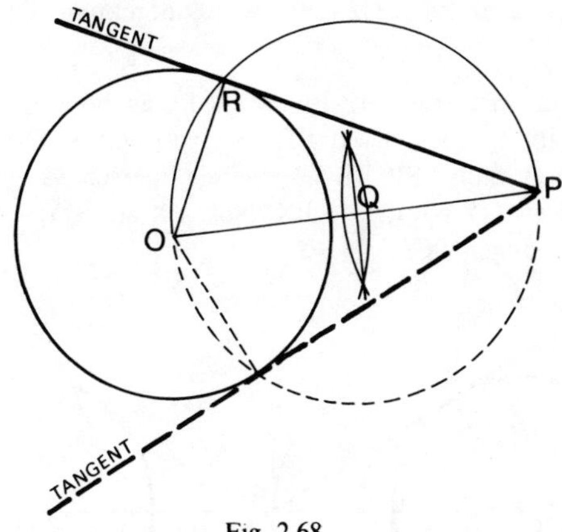

Fig. 2.68

To draw an external tangent to two unequal circles

Example Let the centres of two unequal circles of Ø64 and Ø28 be 109 mm apart. Draw the external tangent.

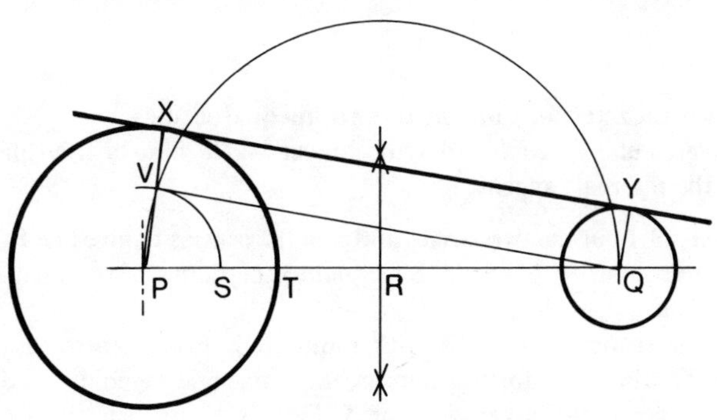

Fig. 2.69

Solution Set out the two circles and join the centres P and Q. Bisect PQ to give point R and draw a semicircle with radius RQ. Mark off ST equal to the radius of the smaller circle. With radius PS, draw an arc from S to intersect the semicircle at V. Draw line PV and extend it to cut the larger circle at X. Draw QY parallel to PX. Join XY to give the required tangent.

To draw an internal tangent to two equal circles

Example Let two circles of Ø64 mm be 109 mm apart. Draw the internal tangent.

Solution Join the circle centres with the line PQ as shown in Fig. 2.70. Bisect PQ to give point R, then bisect PR to give point S. Draw a semicircle from point S with radius SP. This semicircle intersects a given circle at X. Join PX and draw QY parallel to PX. Now PX and QY are both normals to the required tangent XY.

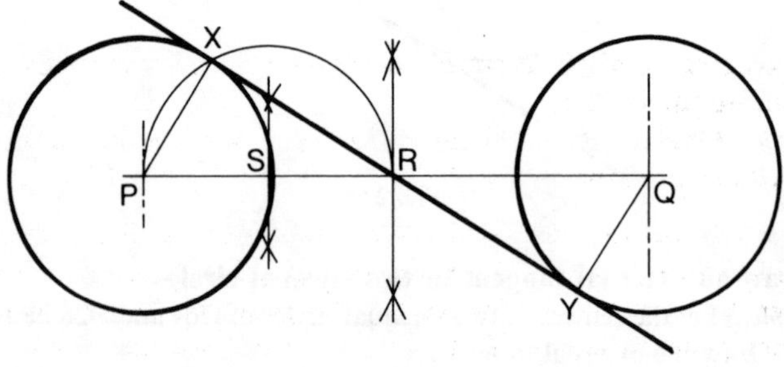

Fig. 2.70

To draw an internal tangent to two unequal circles

Example Let two circles of Ø64 mm and Ø28 mm be 109 mm apart. Draw the internal tangent.

Solution Set out the two circles and join the centres to give line PQ. Bisect PQ to give point R. Mark off ST equal to the radius of the smaller circle, i.e. 14 mm.

Draw a semicircle from R with radius RQ. From centre P, and with radius PT, draw an arc to intersect the semicircle at point W. Join PW which intersects the larger circle at X. Draw QY parallel to PW. The line XY is the required tangent.

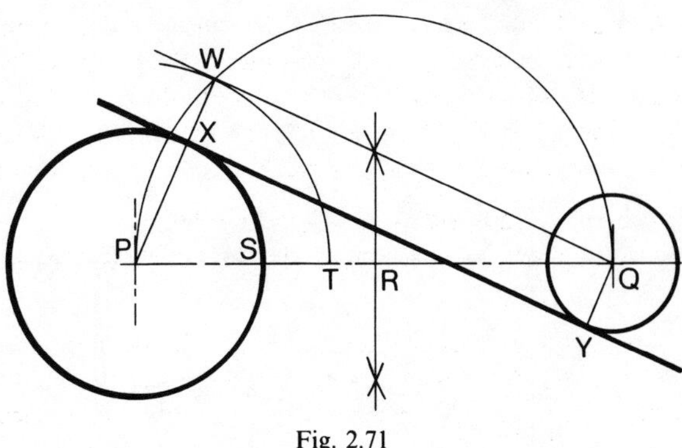

Fig. 2.71

To draw a curve of given radius to touch two circles when the circles are outside the radius

Example Draw an arc of 90 mm radius to touch two circles of Ø80 mm and Ø60 mm as shown in Fig. 2.72.

Solution Set out the circles with centres 100 mm apart. The problem is to establish the centre of the 90 mm radius. When two curves blend together and they are from different centres, then the point of tangency will lie on a line joining the two centres. In this case there are two tangency points T_1 and T_2 along the lines OA and OB. To find point O, draw an arc from A equal to the sum of the two radii R40 and R90 and this arc will intersect with another from point B, equal to R30 plus R90. It is always necessary to determine where curves start and finish so that they may be drawn with precision, especially where the linework is completed in ink. Slight overlaps at points of tangency give a very poor impression.

To draw a curve of given radius to touch two circles when the circles are inside the radius

Example Draw an arc of 120 mm radius to touch two circles of Ø80 mm and Ø60 mm as shown in Fig. 2.73.

Solution Set out the circles as before with centres 100 mm apart. In this case the circle centres, A and B, lie between the centre of the required arc

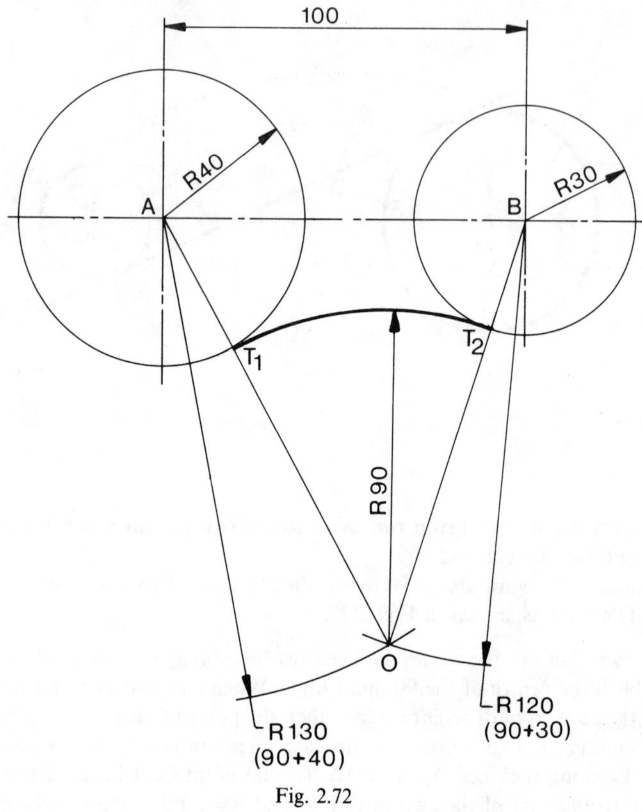

Fig. 2.72

and the tangency points. From point A, describe an arc of R120 minus R40 which gives a radius of 80 mm and this will intersect with another from centre B of R120 minus R30 giving a radius of 90 mm. The intersection establishes the centre of the arc of the required 120 mm radius which can now be lined in after marking the tangency points T_1 and T_2.

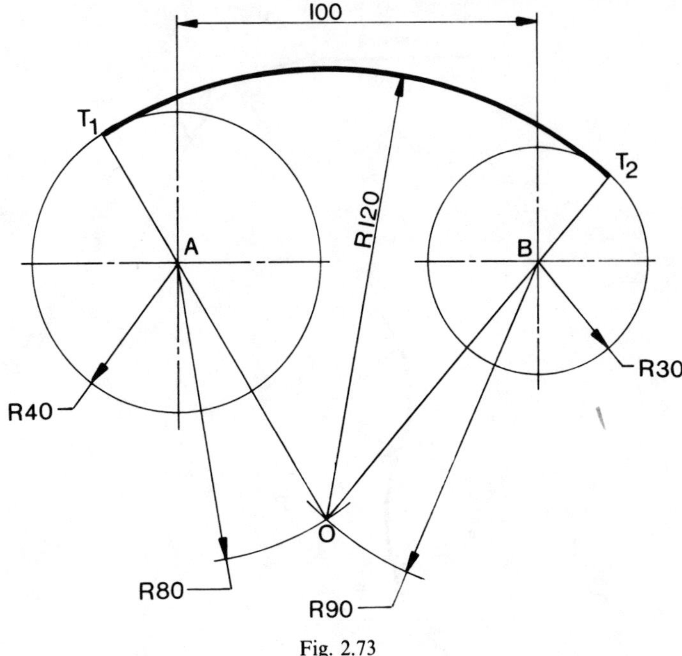

Fig. 2.73

Tangency points

Tangency points shown by the letter P in Fig. 2.74 give the exact position where straight lines join curves and curves join other curves. It is always necessary to establish the location of tangency points to produce neat and accurate linework. Remember that the normal and tangent make an angle of 90° as indicated in Fig. 2.74a, but in Fig. 2.74b the normals from both centres of curvature form the straight line joining the centres.

Fig. 2.75 gives a typical example where arcs blend with straight lines forming acute and obtuse angles. Copy this example and in each corner find the centre of the arc from the intersection of construction lines parallel to the sides of the triangle. The tangency points are fixed by drawing normals from the centres of the corner arcs. All corner radii are 15 mm.

Fig. 2.76 shows a typical example where a corner arc blends with a curve and a straight line. In each of the four corners, the arc centre must be established by a radii from the centre of the boss and its intersection with

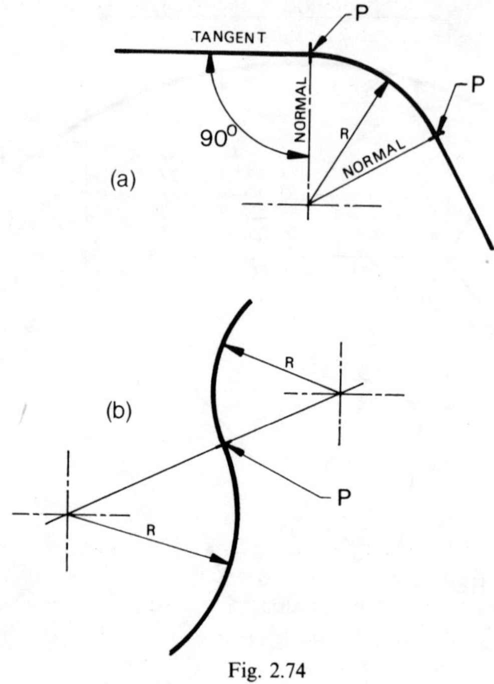

Fig. 2.74

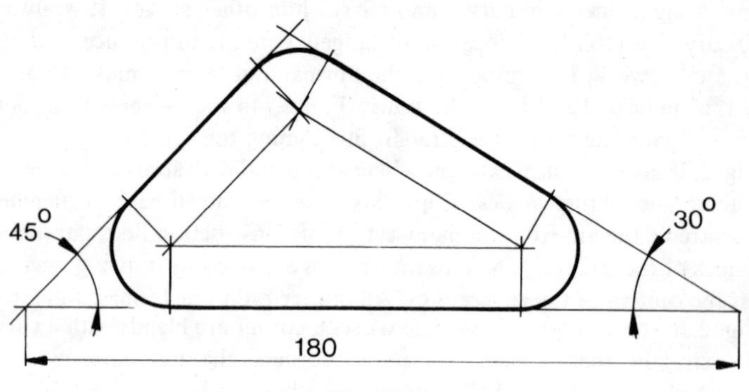

Fig. 2.75

a line parallel to the side. Copy this component drawing and determine the tangency points before lining in the outline.

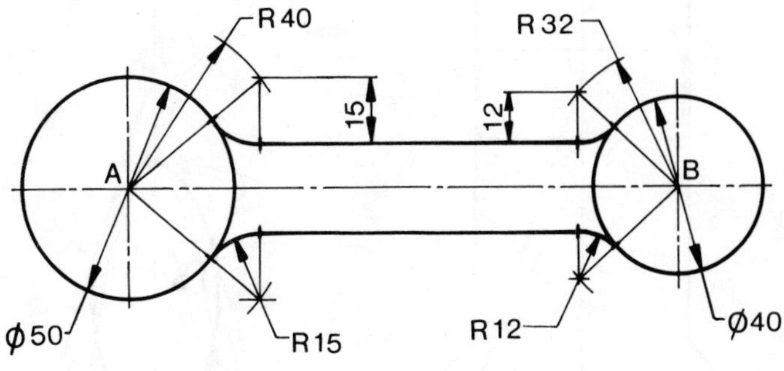

Fig. 2.76

Loci

A *locus* (plural: *loci*) may be described as a straight line, a curve, or a combination of straight lines and curves traced out by a point, line, or surface which moves according to some mathematically defined conditions.

If we take the simplest of examples to illustrate this definition, then the locus of a point P which moves so that it is always the same distance from another fixed point would be a circle. Alternatively, the locus of a point Q which moves so that it is always at a fixed distance from a given straight line, would be a line drawn parallel to it. In order to understand completely the operation of any mechanism, it is necessary to be able to plot the path taken by any of its moving parts. Many motions can be plotted by drawing the new position of a point for selected parts of its travel and then joining the points to give a curve.

Redraw the following problems. In the first example, twelve different positions of the connecting rod are taken by drawing the crank at 30° intervals and the solution gives all of the necessary construction lines.

Example 1 A diagrammatic sketch showing the principle of an internal combustion engine is shown in Fig. 2.77. AB is the connecting rod which rotates on the crank shaft BO and is free to pivot on the piston at A. AB = 84 mm, BO = 30 mm. C is the mid point of AB. Draw, full size, the locus of point C for one revolution of BO.

Middlesex Regional Examining Board

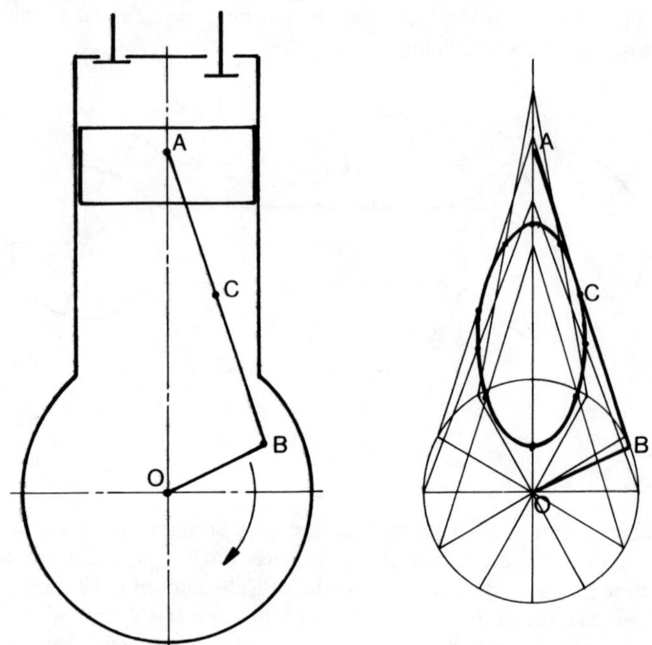

Fig. 2.77 Locus of point C for one revolution of crank.

Example 2 Fig. 2.78 shows a common type of trellis system which can be used to adjust mechanical arms. The trellis pivots about the centre points; point B is a fixed pivot. A and C are operated to give adjustment. Plot the paths of the ends D and E when A and C are brought together to the centreline.

Middlesex Regional Examining Board

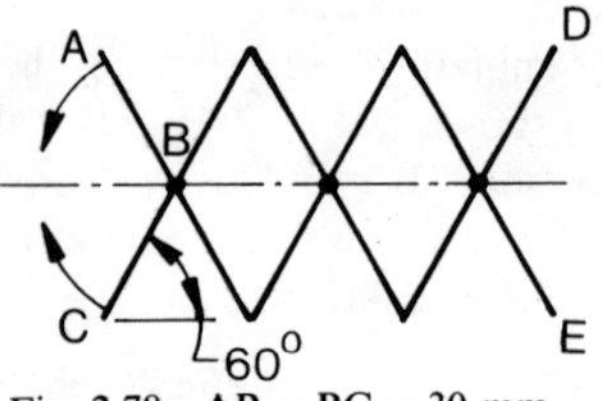

Fig. 2.78 AB = BC = 30 mm

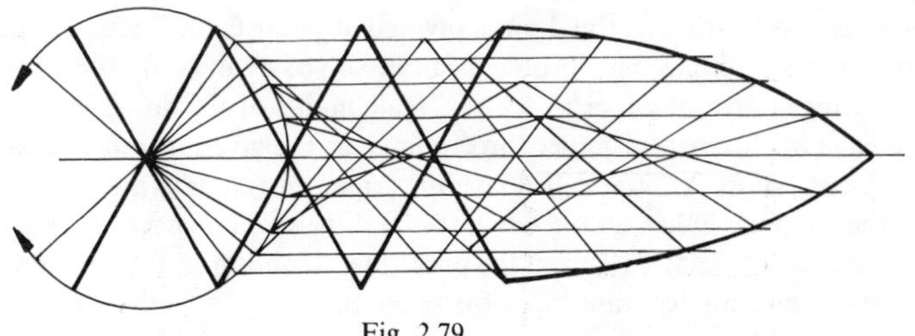

Fig. 2.79

Solution Fig. 2.79. Draw the mechanism in the six alternative positions, shown which do have to be equally spaced, and draw the locus through the different points obtained for the ends D and E.

Example 3 An up-and-over garage door is shown in the partly opened position in the diagram Fig. 2.80. The wheel at the top of the door rolls along a channel from A to A_1 as the door is opened, while the bottom of the door moves upwards from B to B_1. The stay CD pivots on the wall of the garage at C and on the edge of the door at D. The fully opened and closed positions of the door are indicated by the short chain lines.

Fig. 2.81 is a line drawing of the door and mechanism already reduced to scale.

Copy the line drawing to the sizes given and then plot the locus of the bottom of the door as it moves from B to B_1.

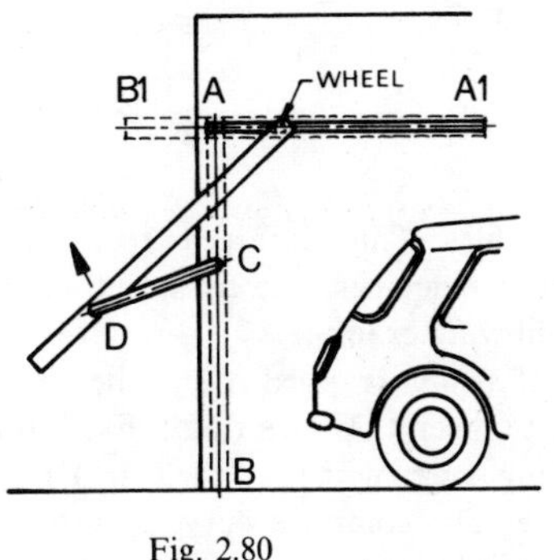

Fig. 2.80

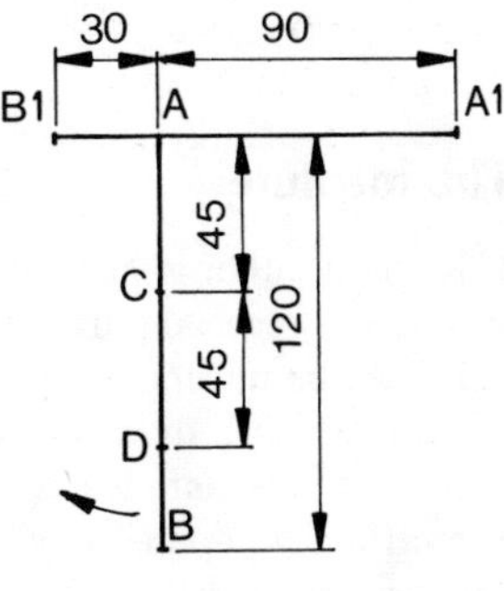

Fig. 2.81

Solution Note that as the door is pivoted at point C, the locus of point D must be a semicircle. This problem can be solved by drawing the semicircle and using a strip of paper to act as a scale model of the door. Mark along the edge of the strip of paper three points, at the given distances apart, to represent points A, D and B. Place the paper on the drawing so that A is on the channel, D lies on the semicircle and then add a pencil mark against the position of point B. Repeat the procedure to obtain sufficient alternative points to give an accurate locus for point B.

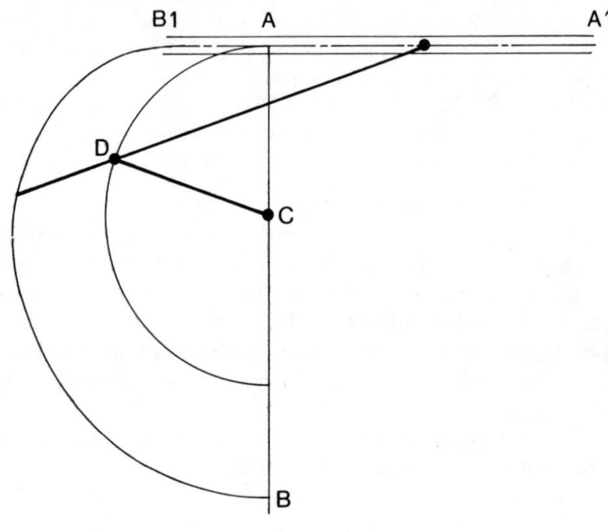

Fig. 2.82

The involute

This spiral curve is applied to the profiles of involute gear teeth and is produced by special purpose gear cutting machines. The construction principles will be understood from the following example.

Assume that a drum has a length of string wrapped round the circumference for one complete turn and that one end of the string is fixed to the drum. If we gradually unwind the string, keeping it taut, then the locus of the free end will describe an involute. The complete curve is shown in Fig. 2.83.

Example Draw an involute curve generated from a base circle of 50 mm diameter.

Solution Draw the given base circle and divide into twelve equal parts. Calculate the length of the base circle circumference, ($\Pi \times D = 3{\cdot}14 \times 50 = 157$ mm) draw a line from point P and divide it into twelve equal parts.

Draw tangential lines from each point around the circumference and wind the string onto the drum by marking along each successive tangent the appropriate length, i.e. from point 11 mark the distance P11, from point 10 mark the distance P10. Draw a curve through the marked points to give the required involute. A line drawn perpendicular to any of these radial lines will give a tangent to the involute where the radial line intersects with the involute. One tangent is shown at point 8.

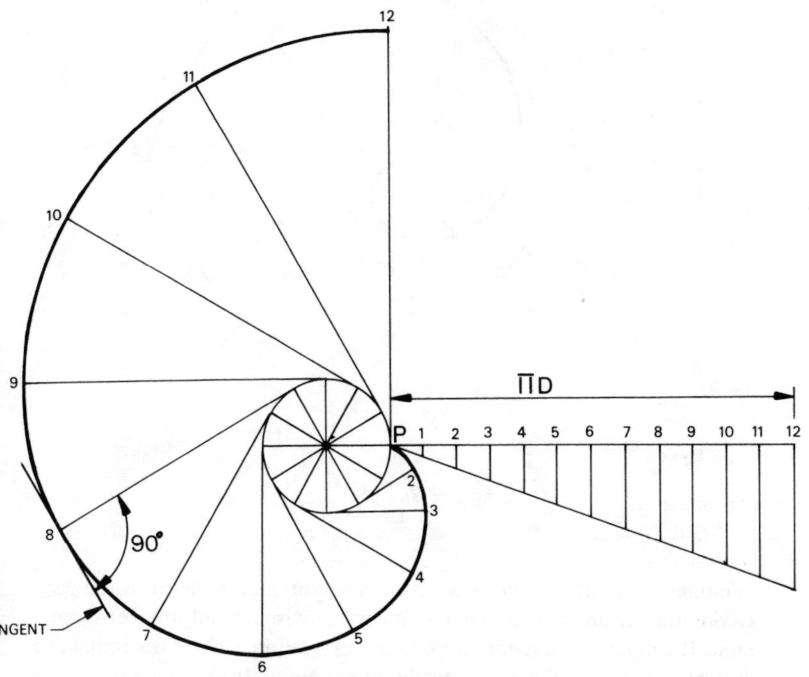

Fig. 2.83

Archimedean spiral

This spiral can be considered as the locus of a point which rotates about a fixed centre at constant speed and at the same time moves away from the centre at constant speed. Curves of this type can be seen in the design of hot plates on electric cookers.

Example Draw an Archimedean spiral for one complete turn, the curve to start at the centre of a circle 140 mm diameter and to finish at the circumference.

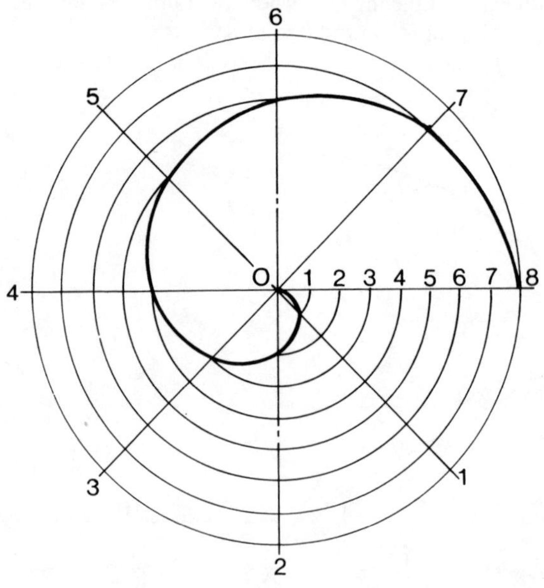

Fig. 2.84

Solution Draw the given circle and divide it into a number of equal parts. Divide the radius into the same number of divisions and number them as indicated. Draw radii from centre O to intersect on each of the radial lines in turn. The required spiral is produced by joining these intersections with a curve.

The helix

The helix is a curve generated on the surface of a cylinder by a point which revolves around the cylinder and at the same time moves up or down the surface. The revolution can take place in a clockwise or anticlockwise direction, giving a right or left hand helix. This direction of rotation can be observed on the plan views in Fig. 2.85 and Fig. 2.86. The lead of the helix is the axial movement of the point, which generates the helix, in one rotation of the cylinder.

Plot the curves shown below by dividing the plan view into twelve parts, divide the lead into the same number of parts and project points around the circumference. All construction lines have been added.

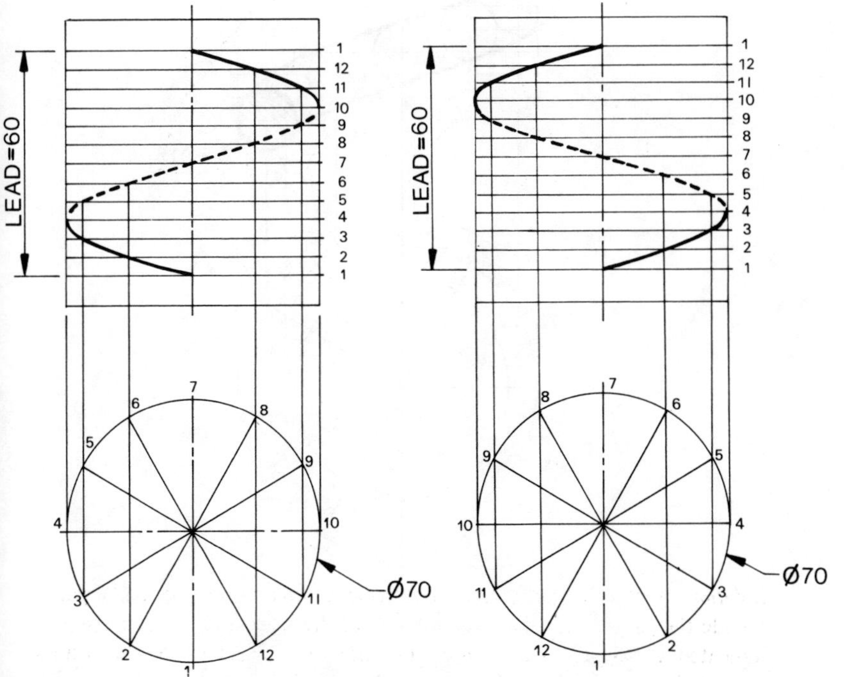

Fig. 2.85 Left hand helix.

Fig. 2.86 Right hand helix.

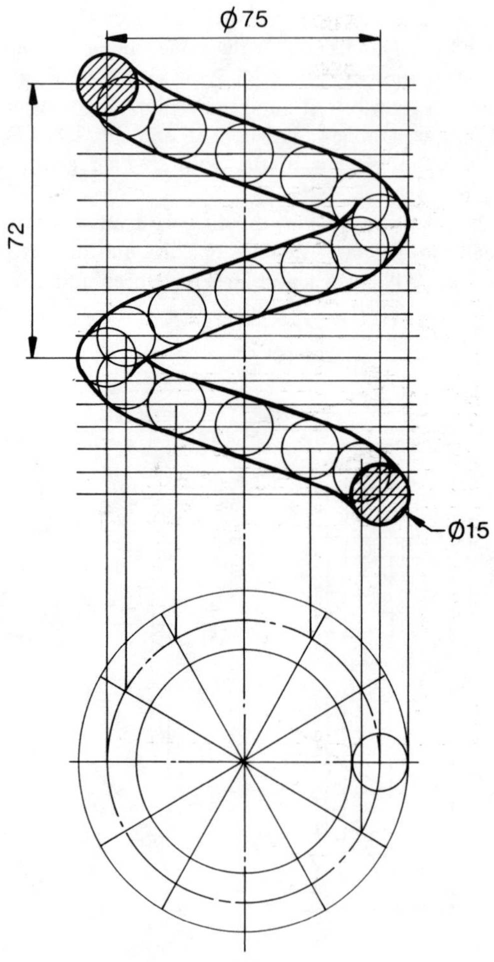

Fig. 2.87

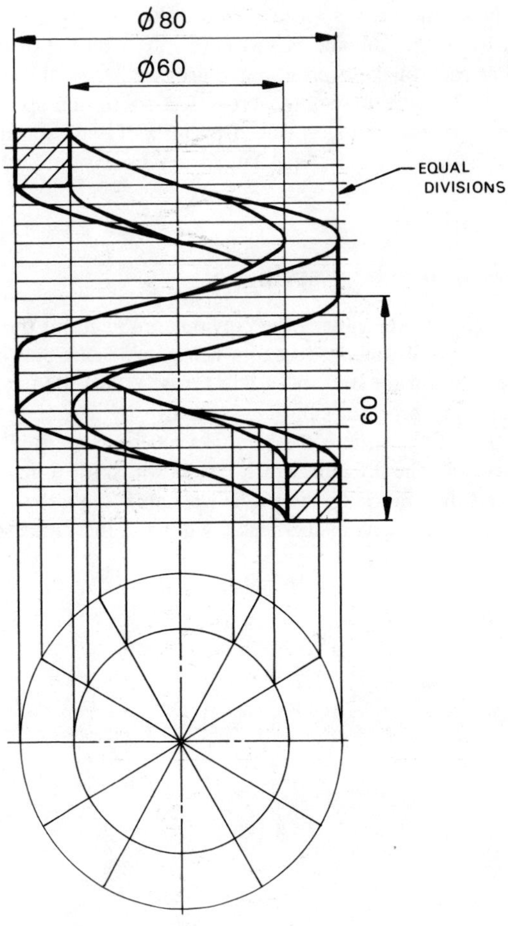

Fig. 2.88

Circular section spring

Fig. 2.87 shows the helix applied to the drawing of a spring. Copy this illustration by setting out the helical path traced by the centreline of the spring and at each of the twelve points, projected from the circumference, draw in a circle to represent the cross section of the spring wire. The outside profile of the spring is traced by lines drawn tangentially on the inside and outside of these circles. All necessary dimensions and construction lines are provided.

Square section helical spring

This example is an extension of the previous problem but the four corners of the square cross section produce four helices which are only partly visible. A complete solution for one and a half turns is illustrated in Fig. 2.88. A second interesting problem can also be drawn by assuming that the top horizontal line of the square cross section has been omitted and that the section is hollow. The drawing which results will show a helical chute.

Redraw this illustration to show a helical chute. Note that as the section is hollow additional parts of the helices will be visible from the corners.

3

Orthographic Projections

Orthographic projection: first angle

An artist will only paint those aspects of an object which he can see from his viewpoint.

Orthographic drawings give views which are produced by observing separate faces of a component in order to demonstrate all its physical features. These separate views are combined on a composite drawing and positioned in relation to one another in a defined pattern.

A simple case is shown in Fig. 3.1 which gives a pictorial view of a block with a groove across the front face and the top left hand corner removed.

Imagine two vertical planes, or surfaces, positioned at right angles to each other and parallel to the sides of the block. An additional horizontal plane is shown beneath the block. Three arrows are shown and the intention is to view each of the faces of the block in turn and project the image on the surface behind each face. The three arrows all lie at right angles to the sides of the block.

The view drawn in the direction of arrow 'A' onto the horizontal plane will give an illustration of the top of the block and this view is generally known as a plan view. Although we cannot see the groove through the block, its position is indicated by the dotted line.

The front view in the direction of arrow 'B' is often referred to as a front elevation. The projected view in line with arrow 'C' gives a side or end view and this is also often described as a side or end elevation.

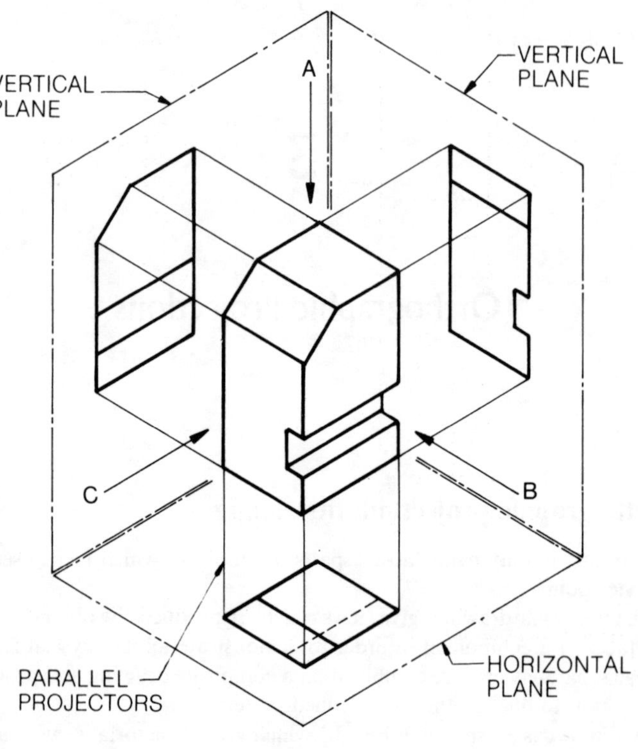

Fig. 3.1

A draughtsman would present these three views together as an orthographic drawing as shown in Fig. 3.2. Note that the separate views are always projected in line with one another. The spaces between each view can accommodate dimensions and notes, so the widths of these spaces are left to the discretion of the draughtsman. The construction lines have not been erased in this illustration to emphasize the fact that the views are in line.

There are several points which should be memorised in connection with this example of first angle projection:

(*a*) Heights in the front and side views are the same.

(*b*) Widths in the side view are equal to depths in the plan view.

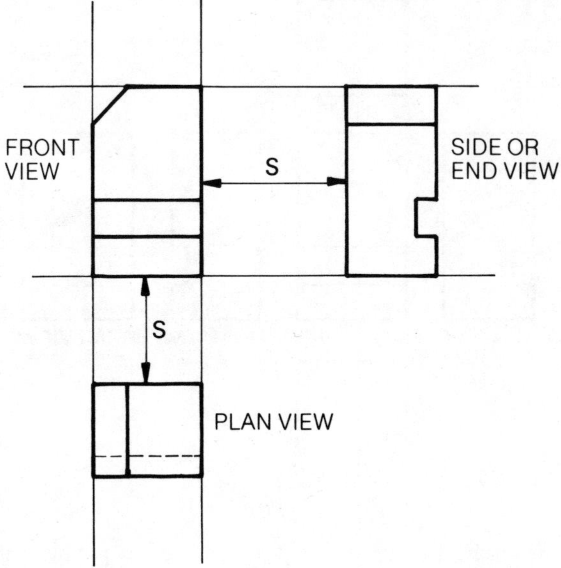

Fig. 3.2 Distance S indicates space required to separate views and add revelant dimensions.

(*c*) The side view, drawn at the right hand side of the front view, actually shows the left hand face of the component.

(*d*) The plan view, which is drawn beneath the front view, shows the top face of the component.

Note that the component could have been totally enclosed on all six sides by the addition of two vertical, and another horizontal plane. This means that six separate orthographic views are possible and these are all illustrated in Fig. 3.3. To appreciate the orientation of the separate views, it is sometimes helpful to use a model and this simple block has been chosen as a model can easily be made from the polystyrene foam which is commonly found in packing. The material can be shaped with a sharp penknife. The three additional views in Fig. 3.3 are clearly obtained by viewing from the right hand side, the rear and from beneath the block.

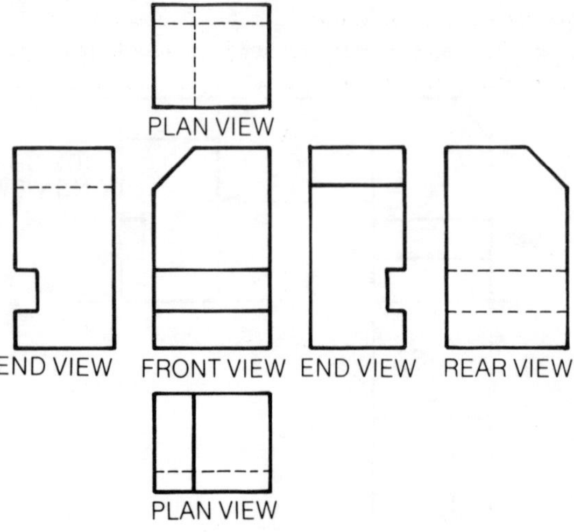

Fig. 3.3

Quite obviously, a draughtsman does not need to produce all of these views to describe adequately a simple object and the views presented would be left to his discretion. It is customary to draw the minimum number of views which illustrate the article completely. It is also a fact that many components do not have a clearly defined 'front', 'back' and 'side' and so the orientation would also be decided by the draughtsman who tries to position the part to convey the maximum amount of visual information.

It should also be pointed out that the terms 'front view' and 'front elevation' should be regarded as having a similar meaning. Drawing standards for engineering refer to 'views' but builders and civil engineers refer to 'elevations'. We also of course refer to 'house plans' and these consist of elevations and plan views and usually the drawing for a house contains more elevations than plans. The term 'bird's eye view' will be associated with the plan view and is accepted to be a view which looks vertically downwards. Conversely, a plan view can also be drawn by looking vertically upwards and this is sometimes called an 'inverted plan'. A little flexibility is needed when interpreting drawings in this respect.

Orthographic projection: third angle

Assume in this alternative method that we have positioned the block behind three transparent surfaces as shown in Fig. 3.4, and draw with a felt pen on each surface the projected view obtained from looking in the directions of arrows A, B and C. The arrows lie at right angles to each surface.

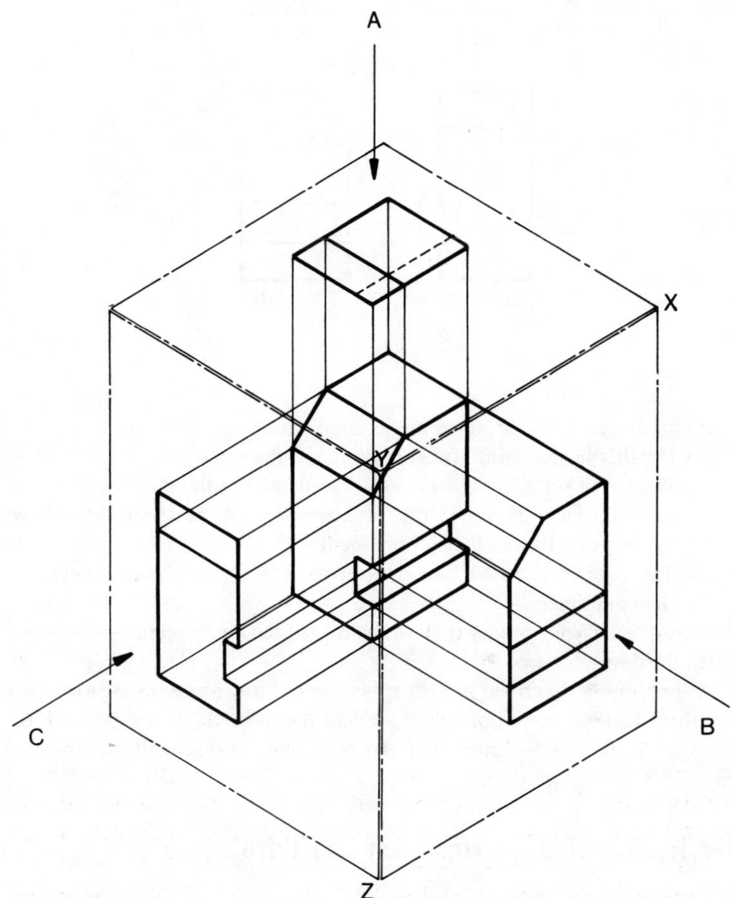

Fig. 3.4 Block positioned parallel to three transparent surfaces.

The surfaces can now be rotated so that the three views are arranged as illustrated in Fig. 3.5, and this is a typical presentation in third angle projection.

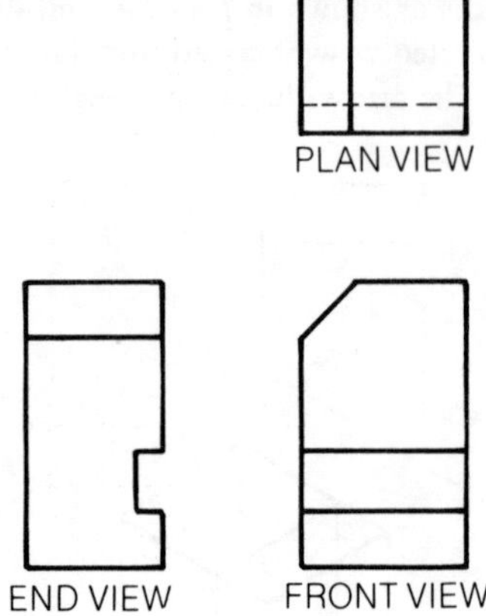

Fig. 3.5

Points to remember in third angle projection:

(*a*) Heights in the front and end views are the same.

(*b*) Widths in the end view are equal to depths in the plan view.

(*c*) The end view drawn at the left hand side of the front view shows the left hand side of the component.

(*d*) The plan view drawn above the front view shows the top surface of the component.

Check using your model that you can turn the component to give these three third angle views.

As previously described in the explanation of first angle projection, there are three further views possible if we had observed the block through the sides of a transparent cube and the complete arrangement is shown in Fig. 3.6.

Derivation of the terms 'first and third angle'

The designations 'first and third angle' can be compared with conventional mathematical practice if we look in the direction of the arrow 'A' in Fig. 3.7. The horizontal plane can be regarded as the 'X' axis and a vertical plane as the 'Y' axis. In first angle projection we project onto the three planes as illustrated. For third angle projection the object is viewed through the planes as previously explained in the description of Fig. 3.4.

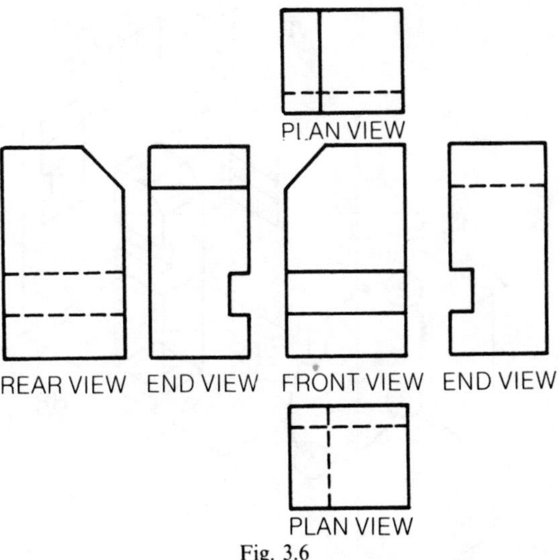

Fig. 3.6

With first angle projection the object is positioned in the first quadrant. Similarly, in third angle the object is placed in the third quadrant. Remember that in each quadrant in order to completely surround the component we can have a total of four vertical and two horizontal planes giving six possible views but that generally three are usually adequate to illustrate any particular solid.

Drawing symbols to indicate projection method

The current British Standard makes the point that both first and third angle projections are equally acceptable and approved internationally. It is a fact however that first angle is the projection method most widely used throughout Europe and third angle is the system used in North America. To remove all doubt, a distinctive symbol should be added on each completed drawing to assist the reader and these are indicated below. Many drawing offices use drawing sheets which are printed with a prepared format and projection symbols are included. The symbols are drawings of tapered cylinders and note the way the cylinders are turned to obtain the required views.

The current International Standards Organisation publication R128 contains the following note, 'For uniformity among the figures in this Inter-

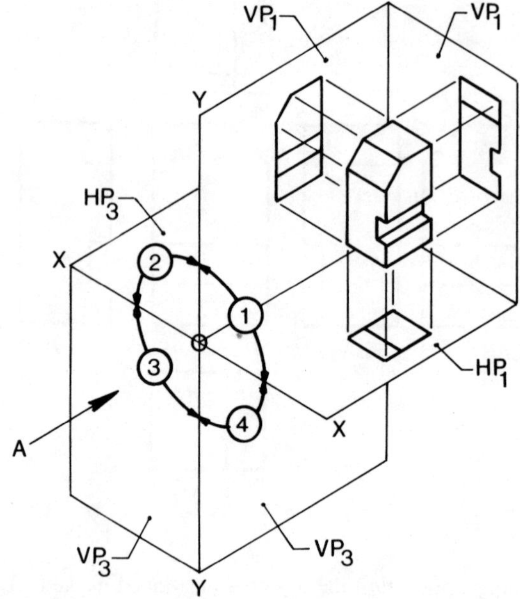

Fig. 3.7 Vertical and horizontal planes used for first angle projection are shown as VP_1 and HP_1.

Vertical and horizontal planes used for third angle projection are shown as VP_3 and HP_3.

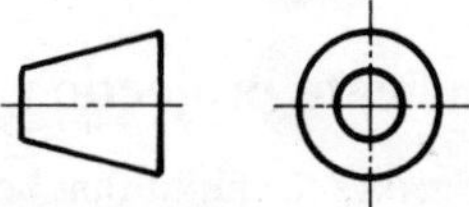

Fig. 3.8 First angle or European projection symbol.

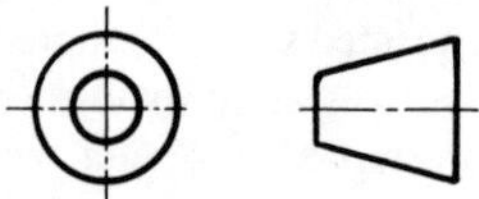

Fig. 3.9 Third angle or American projection symbol.

national Standard, the relative positions of views are those provided by the first angle projection method. It should be understood that alternative projection methods could well have been used without prejudice to the principles established'. In this book we have followed this recommendation and unless otherwise stated the illustrations have been prepared in first angle projection.

First angle projection exercises

Fig. 3.10 shows pictorial views of eight solid components presented on grids formed by equilateral triangles with sides of 10 mm. All dimensions are in increments of 5 mm or 10 mm.

In all cases draw a front view in line with the given arrow and project a plan view beneath. With problems 1 and 7, project an end view on the left side of the front view. For problems 2, 3, 4, 5, 6 and 8, project the end view on the right side of the front view. Include all dotted detail.

The solutions to this and the following three problems are on pp. 94–100.

First angle projection comprehension exercises

As a test of comprehension, make freehand pictorial sketches of the eight solid components shown in Fig. 3.11. The drawings were prepared on a square grid with 10 mm sides. Good proportion is an essential part of a sound sketching technique so try to keep the sketches approximately full size.

Third angle projection exercises

Fig. 3.12 shows pictorial views of four solid objects which are presented on a grid formed by equilateral triangles with sides of 10 mm length. In each case, sketch a front view in the direction of the arrow, then project a plan view above and an end view on the right hand side of the front view. Make your sketches approximately full size.

Third angle projection comprehension examples

Fig. 3.13 shows third angle projections of four components. As a test of comprehension, sketch a pictorial view of each component and place the corner marked 'A' in the foreground of the solution. The drawings were prepared on a square grid with sides 10 mm in length. Take particular care with the proportions of the sketches and try to draw them approximately full size.

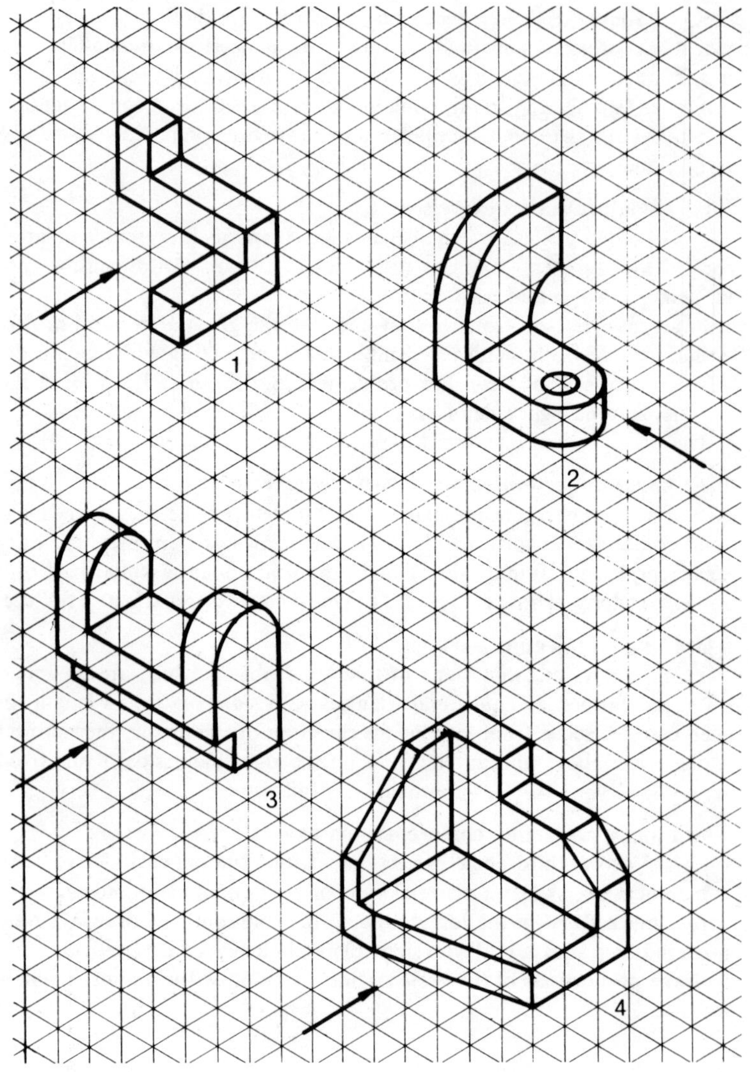

Fig. 3.10 First angle projection exercises.

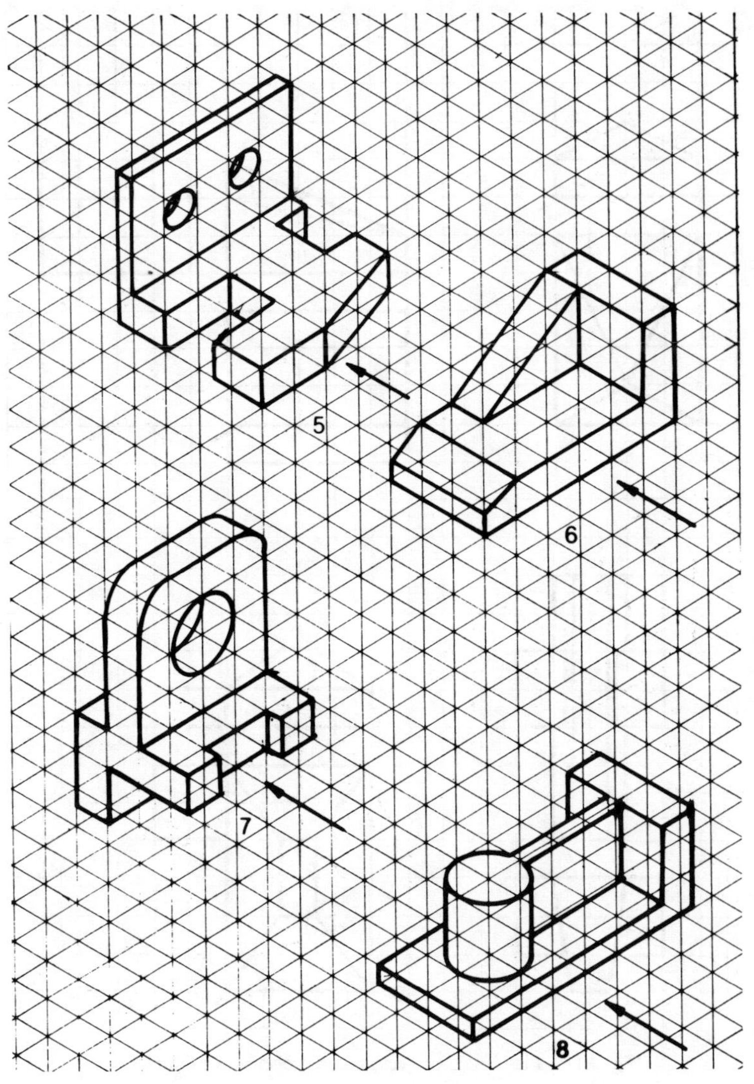
5
6
7
8

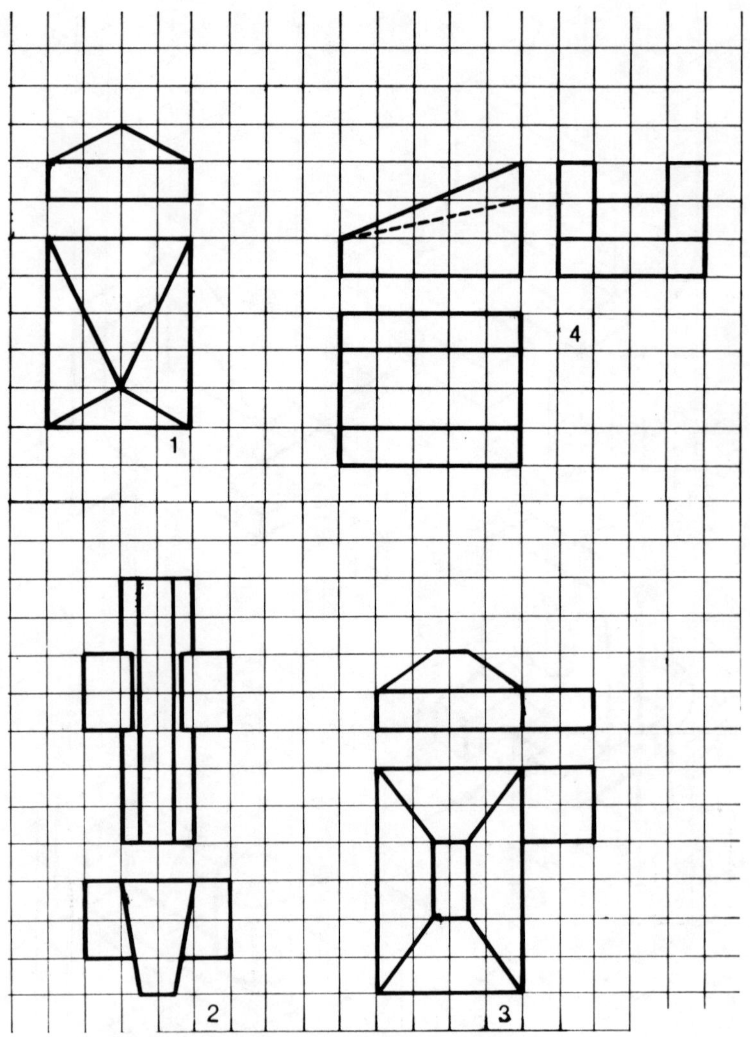

Fig. 3.11 First angle projection comprehension examples.

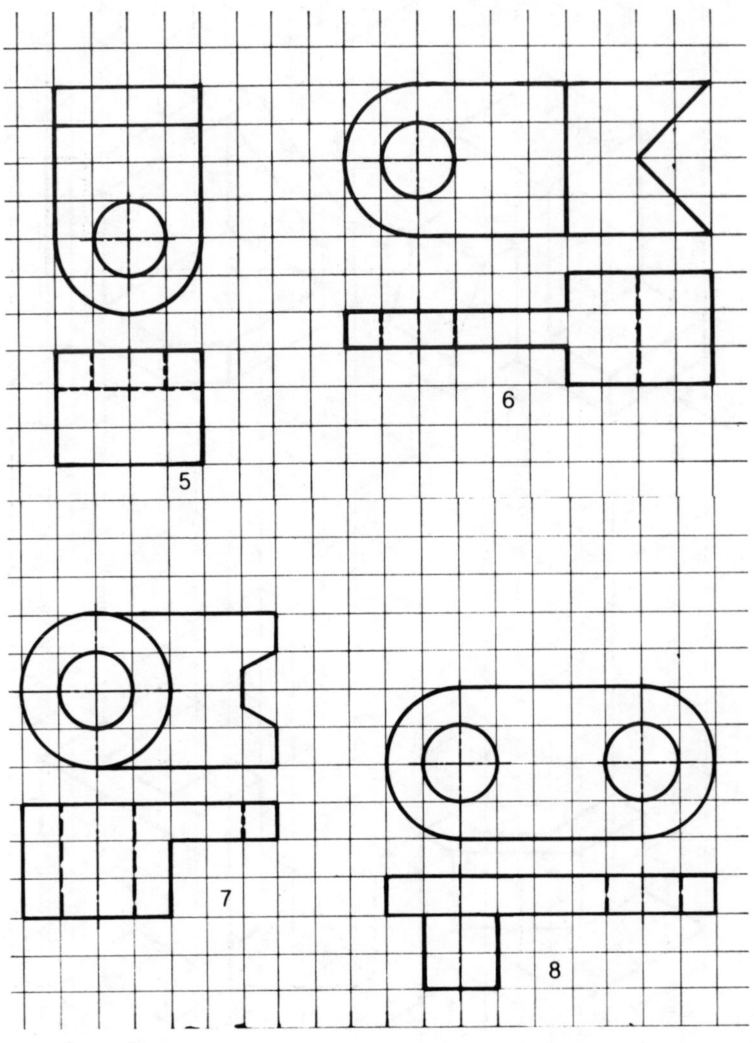
5
6
7
8

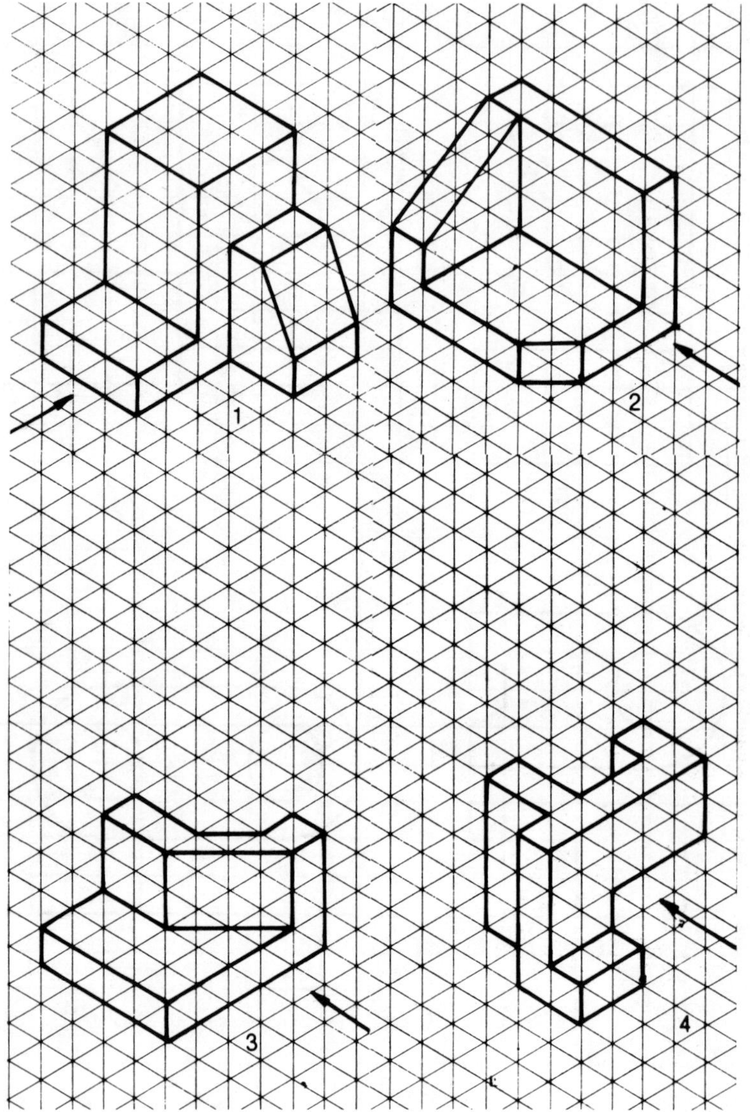

Fig. 3.12 Third angle projection exercises.

The selection of particular views

The draughtsman must use his discretion regarding the presentation of sufficient views to describe adequately a given object.

All orthographic drawings require at least two views to convey the three dimensions of length, breadth and width on two dimensional paper. Alternatively, one view must be given with additional notes in the simplest of cases. The drawing of a single circle could represent a ball bearing or the end view of a cylinder, so further information must be supplied. A single line on a drawing can be interpreted as an edge, or a change in direction of a surface, so the reader must glance at two views simultaneously to determine the correct choice. The draughtsman has the added responsibility of ensuring that no possible doubt or ambiguity exists when his drawing is being read by the user, and this can only happen if he has selected an adequate number of correctly oriented views.

Fig. 3.14 shows a number of cases where the front views are similar but the plan views are different.

In Fig. 3.15 the reverse situation is illustrated with similar plans and alternative front views. A little imagination is necessary when reading technical drawings to visualise the substance of the drawing in three dimensions. With very complicated components and assemblies, it may be necessary from the appreciation point of view for the draughtsman to trace additional sections and auxiliary views, these are dealt with in a later chapter.

Four simple solids are illustrated in Fig. 3.16 to emphasise the case where front and plan views are identical. Note that in each component the end view is essential to define the outline. The plan and front views provide information regarding the thickness of the part, but as they are alike one of them is redundant. We would reject the plan and keep the front view. A finished drawing for each of these examples would give only the front and side views.

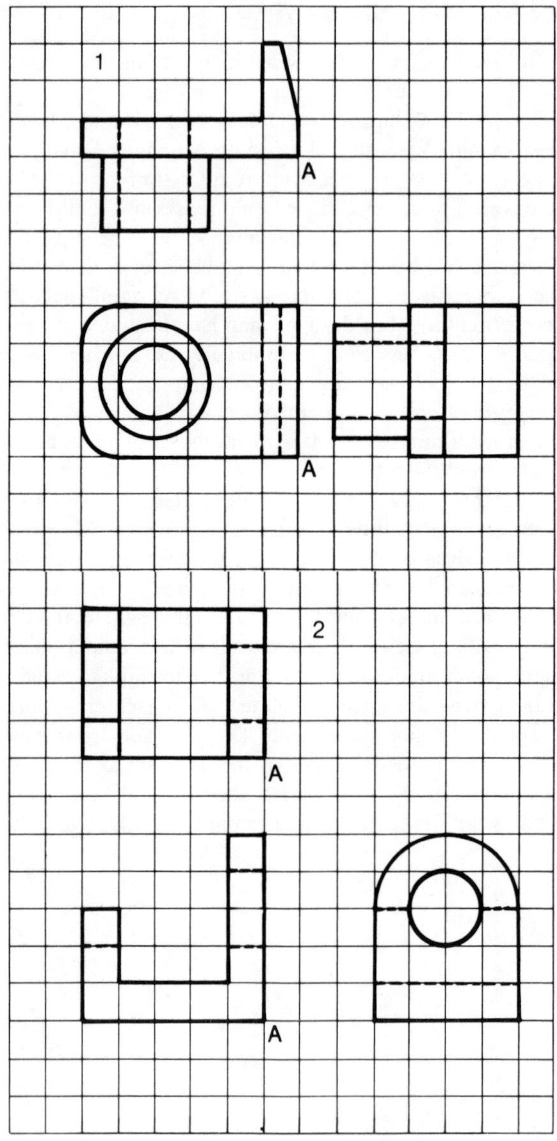

Fig. 3.13 Third angle projection comprehension exercises.

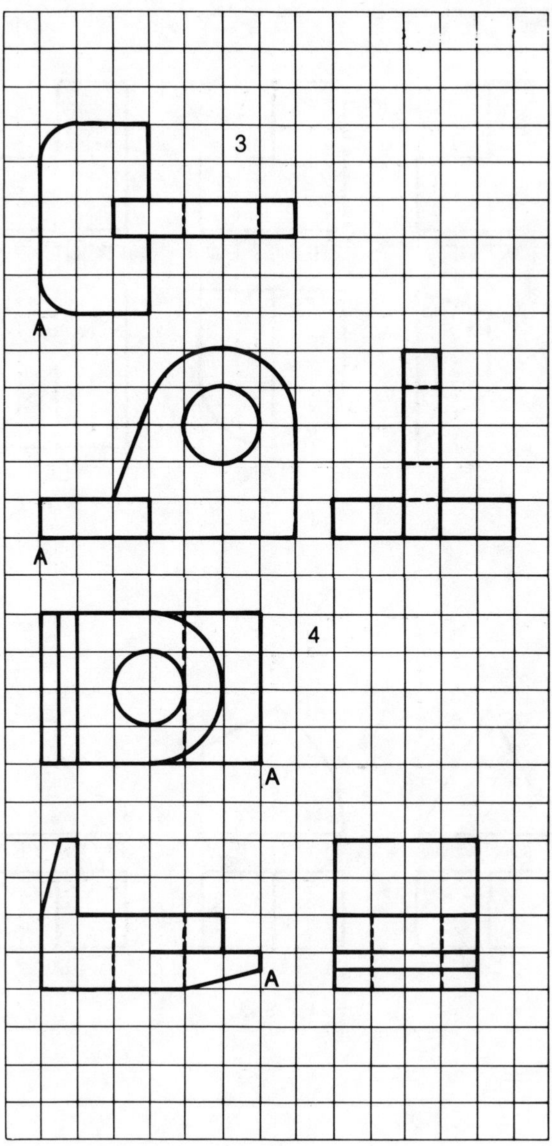
3
A
A
4
A
A

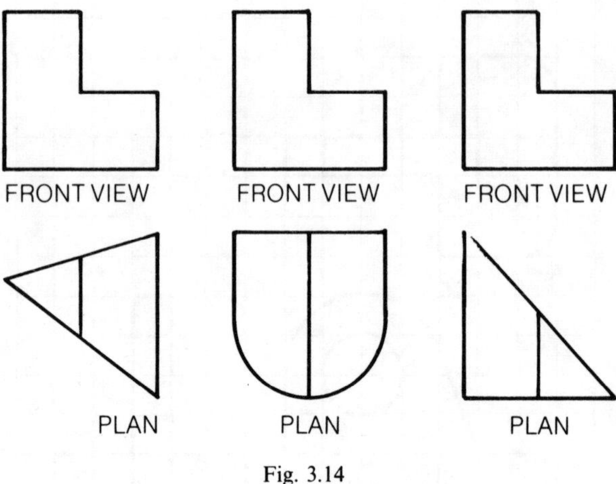

Fig. 3.14

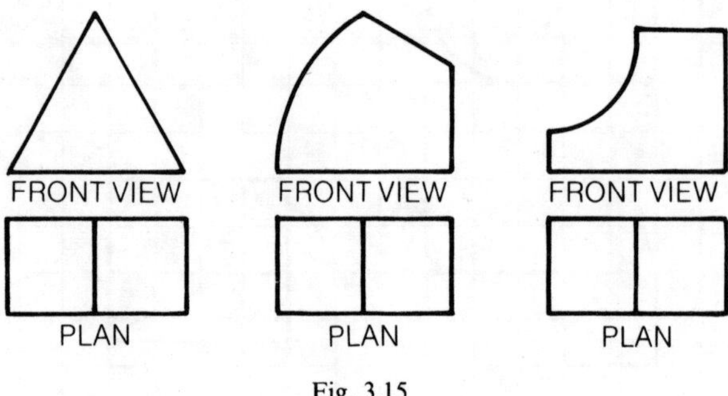

Fig. 3.15

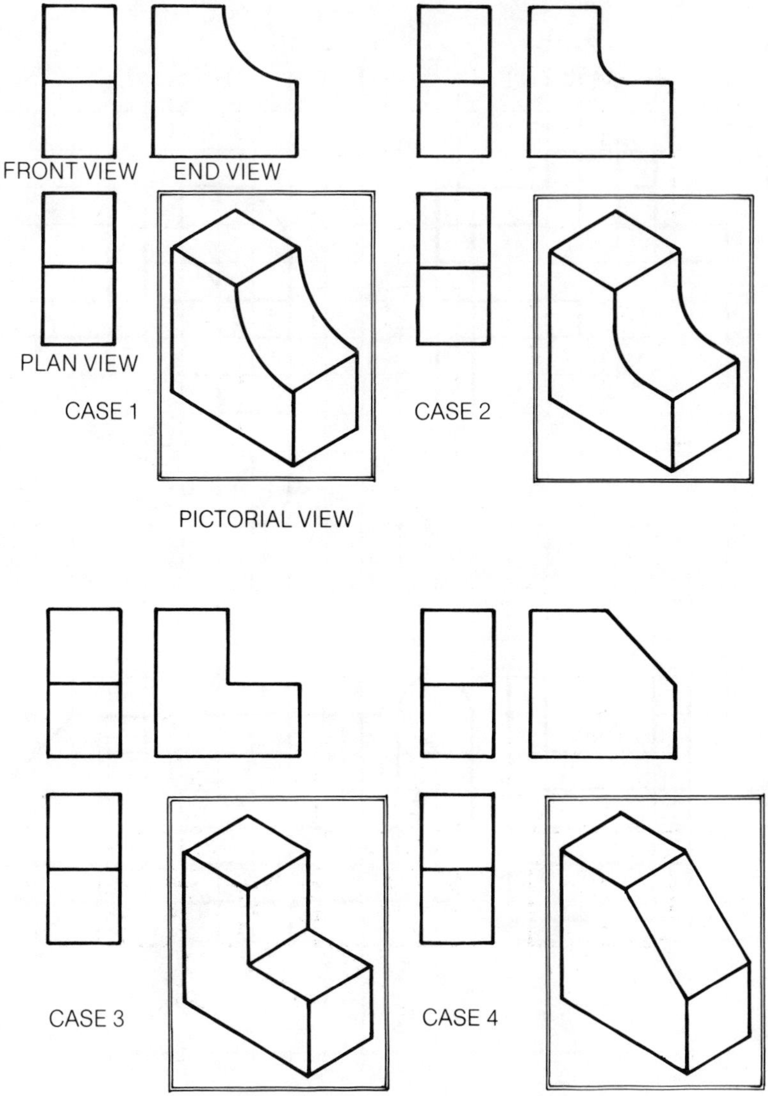

FRONT VIEW

END VIEW

PLAN VIEW

CASE 1

PICTORIAL VIEW

CASE 2

CASE 3

CASE 4

Fig. 3.16

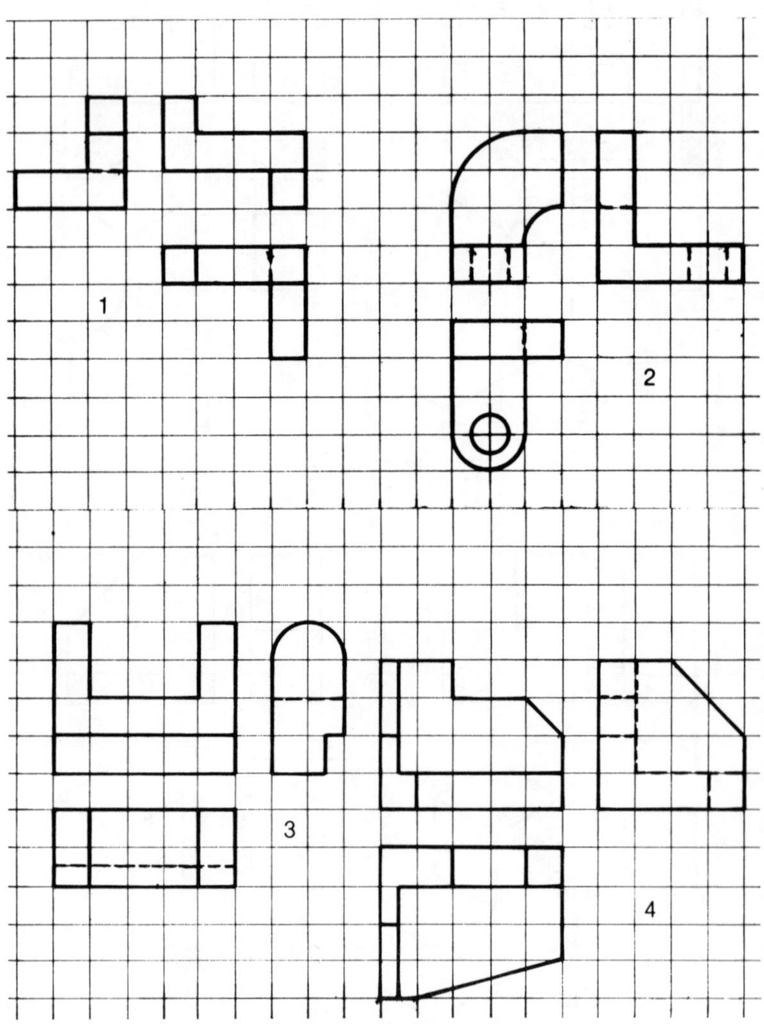

Fig. 3.17 Solutions to first angle projection exercises.

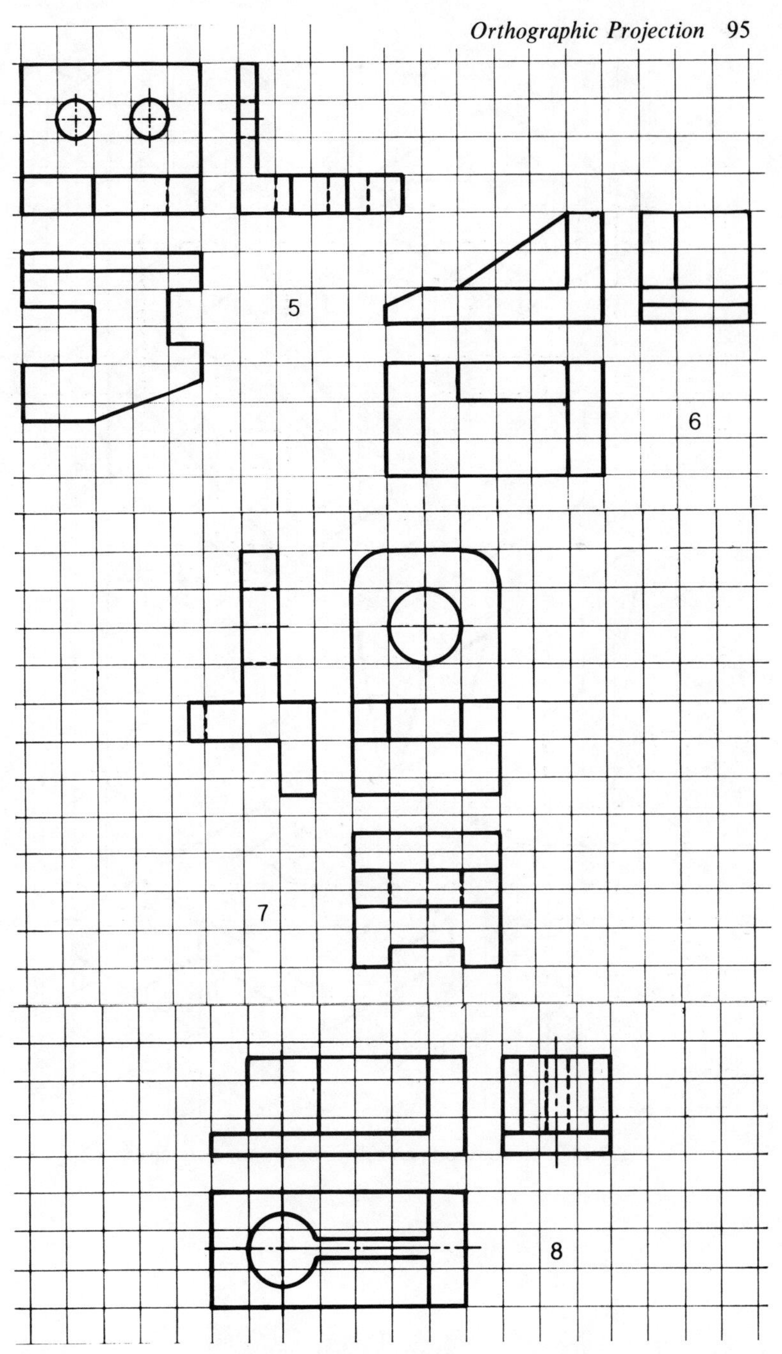

5
6
7
8

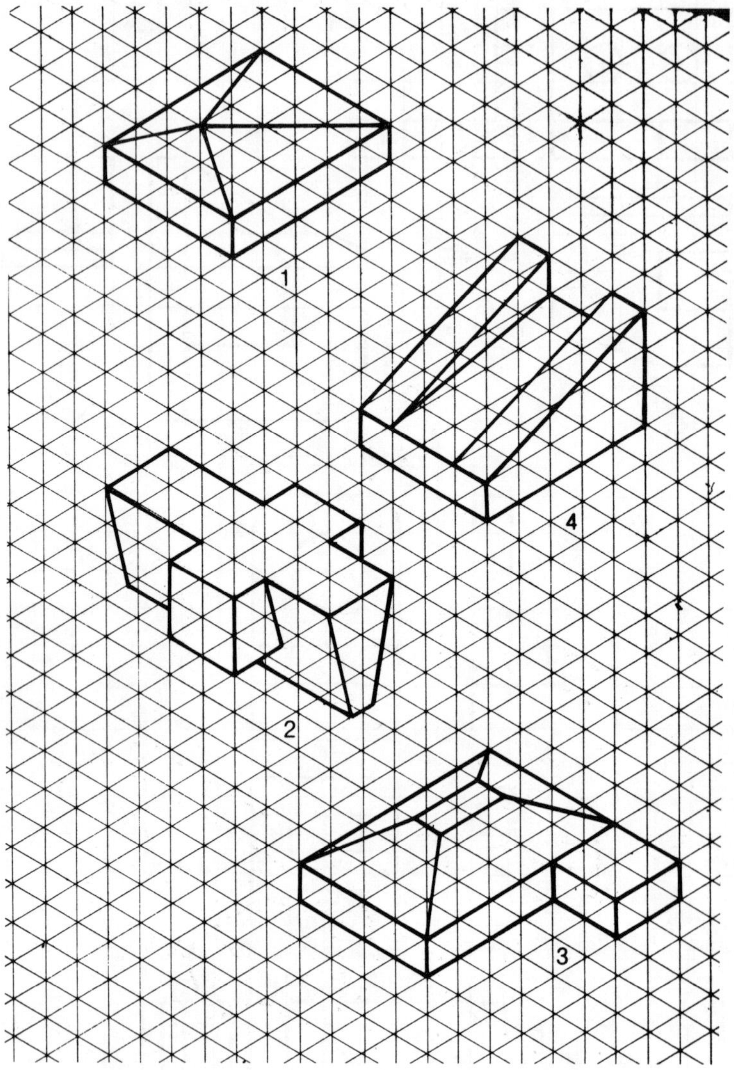

Fig. 3.18 Solutions to first angle comprehension examples.

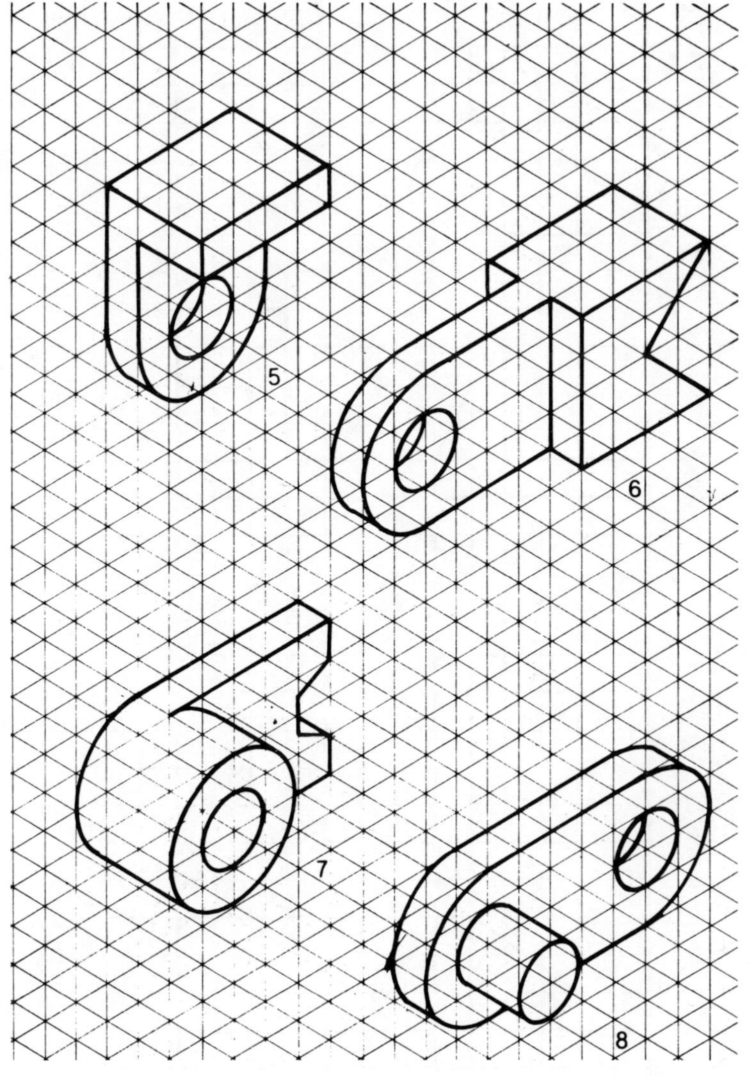
5
6
7
8

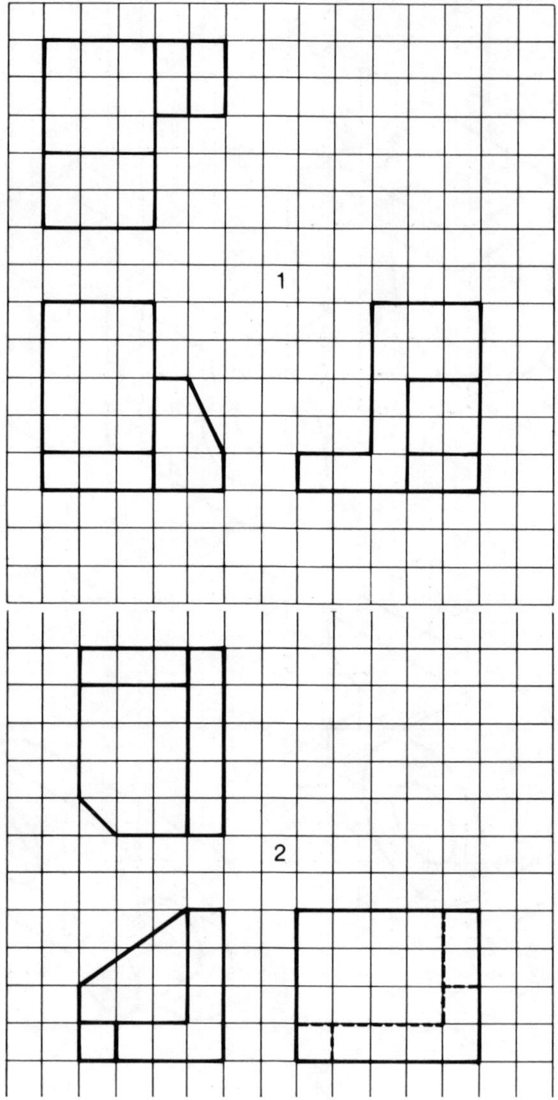

Fig. 3.19 Solutions to third angle projection exercises.

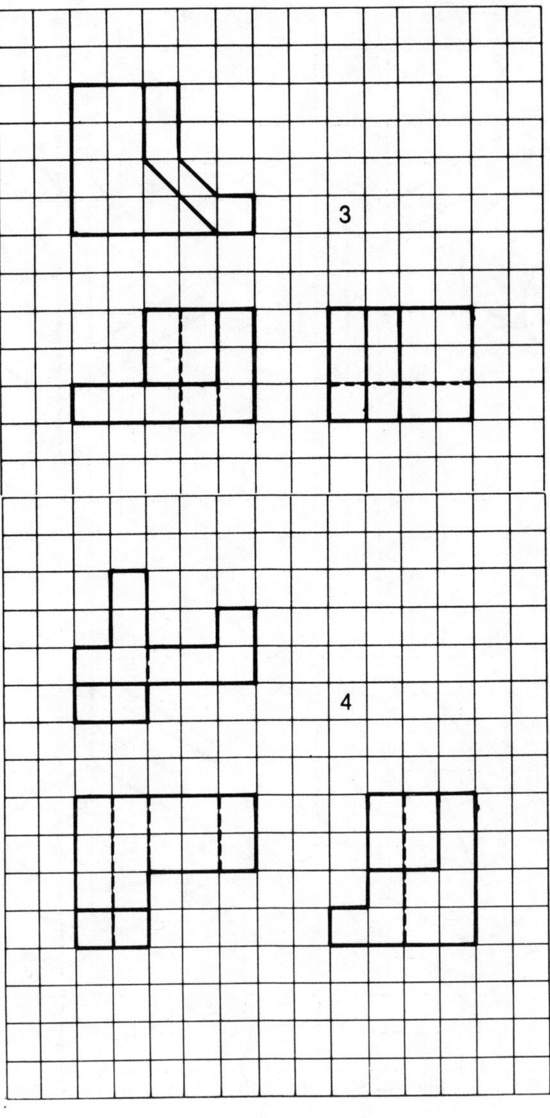
3
4

Fig. 3.20 Solutions to third angle comprehension examples.

4

Applications of Solid Geometry including the Ellipse, Auxiliary Projections and Sections through Solids

The ellipse

Fig. 4.1 shows a piece of circular section bar which has been cut at an angle. Draw the given front view and plan to the dimensions provided. The projection of the face shown as AB, in the end view, will appear as an ellipse. This important curve can be plotted by a simple construction. Take any point along the sloping surface in the front view and project vertically to the plan and horizontally to the proposed end view. The width across the ellipse in the end view at this position is equal to the chordal width in the plan and is shown as dimension 'X' between P_1 and P_2. Repeat this construction for more points along the sloping face AB and line in the required ellipse. The maximum and minimum dimensions of the ellipse are known as the major and minor axes.

There are several other ways of drawing ellipses and for a clear understanding of the methods we would suggest the following examples.

Concentric circle method

Example Draw an ellipse with major axis = 140 mm and minor axis = 100 mm.

Solution Draw the major and minor axes AB and CD which intersect at point O. With centre O, draw two concentric circles. Divide these circles into a number of parts by drawing radial lines through the centre and where

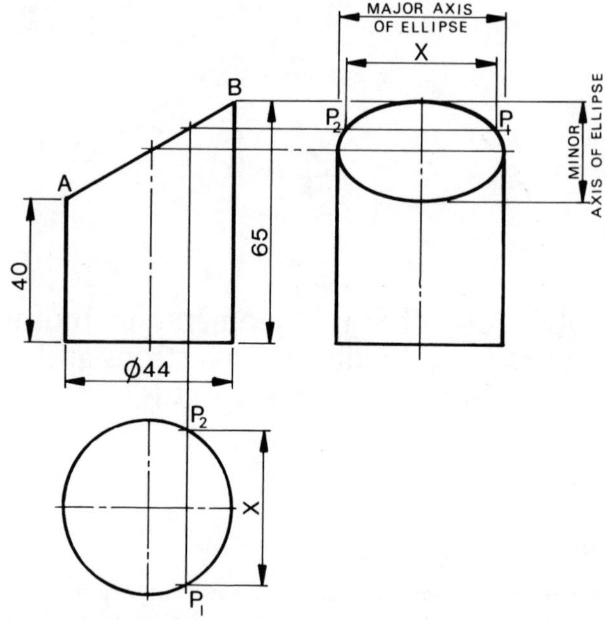

Fig. 4.1

these radial lines cross the major circle drop perpendicular lines to intersect with horizontal lines drawn from the points where the radials cross the minor circles. The resulting intersections give points on the required ellipse.

To draw a normal and a tangent to any point on an ellipse

Draw any ellipse and let the major and minor diameters be AB and CD. The diameters intersect at O. With centre C, and radius equal to half of the major diameter, describe an arc to intersect the major diameter at F_1 and F_2 which are known as focal points. Let P be any point on the ellipse where a tangent or normal are required. Bisect angle F_1PF_2 to give the normal at P and erect a perpendicular at this point, with the normal as the base, and this line will be the tangent.

Now the ellipse is the locus of a point which moves so that the sum of its distances from the two focal points is a constant. In Fig. 4.3, let P_1 be any other point on the ellipse. Measure the distances F_1PF_2, also $F_1P_1F_2$, and they will be found to be the same and the lengths of each are equal to length AB which is the constant referred to above.

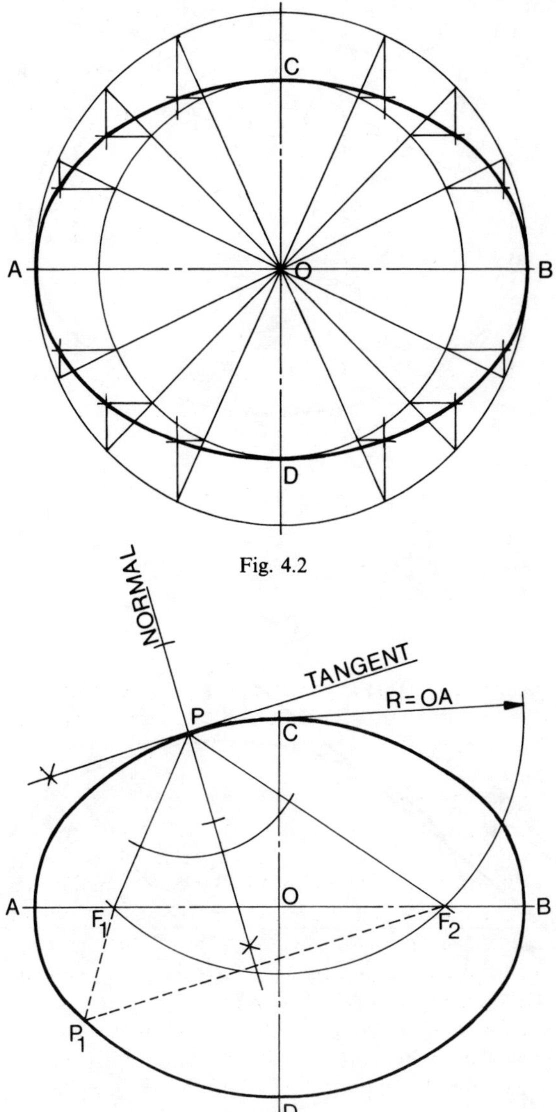

Fig. 4.2

Fig. 4.3

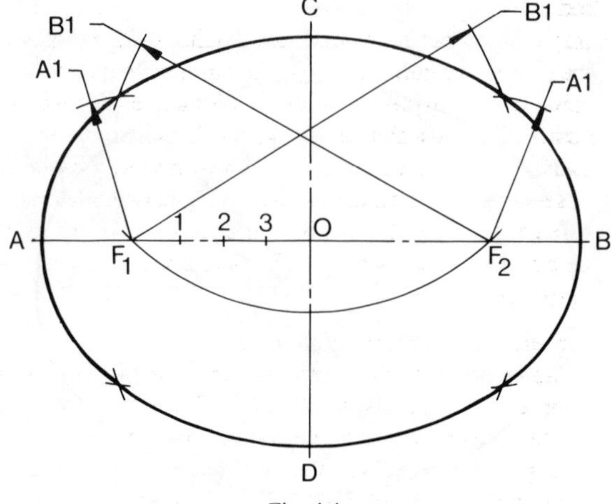
C
B1
B1
A1
A1
A
1
2
3
O
B
F1
F2
D

Fig. 4.4

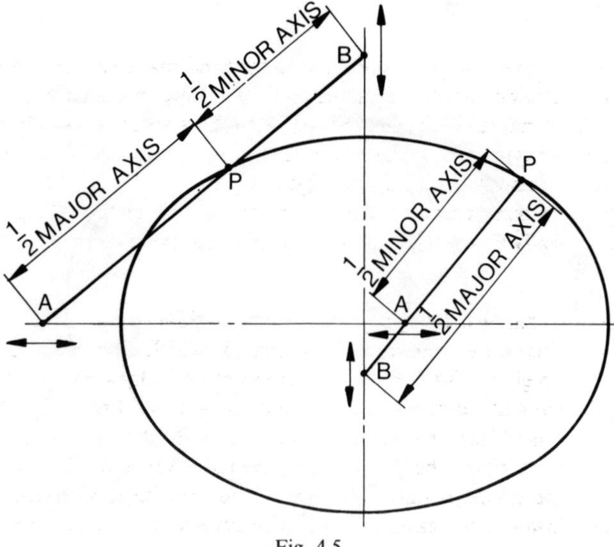
1/2 MINOR AXIS
B
1/2 MAJOR AXIS
P
A
1/2 MINOR AXIS
P
1/2 MAJOR AXIS
A
B

Fig. 4.5

Foci method

An ellipse may be constructed from the focal point using the above principle.

Set out the major and minor axes and determine the position of the focal points as previously explained. Divide the distance F_1O into a convenient number of parts. Four are shown in Fig. 4.4. Describe an arc from centre F_1 with a radius equal to the length A1 above and below the major axis. Let these arcs intersect with radii drawn from F_2 equal to length B1. Repeat this procedure from F_2 and F_1 to give a total of four points of intersection which all lie on the ellipse. Additional points are constructed using radii A2 and B2, also A3 and B3.

Drawing the ellipse by trammel methods

Method 1 Along the edge of a strip of paper, mark half the major and half the minor axes in line, shown here at AP and BP. Position the strip (known as a *trammel*) on the drawing so that point A lies on the horizonal centreline and point B lies along the vertical centreline. There are many positions that can be taken and if a pencil mark is placed against alternative positions of point P and a line is drawn through them an ellipse will be produced. The illustration above shows the situation for a quarter of the ellipse so the trammel will have to be turned round for the other parts. A little experimentation will enable you to draw perfect ellipses by this method.

Method 2 An alternative method uses a trammel where half the major axis and half the minor axis is measured from the same end and this is shown on the right hand side in the illustration above. You will find that this method is satisfactory where the distance between A and B is reasonably large. Mark the path of point P to give the required ellipse.

In both of these methods shown, the major axis of the ellipse will be drawn vertically if point A on the trammel is positioned along the vertical centreline.

Ellipse construction by approximate arc method

Set out the major and minor axes AB and CD which intersect at O. Draw radius OA to give point E on the extension of OC. Draw an arc of length CE to intersect at point G on a line joining A to C. Bisect the distance AG so that the bisector cuts the major axis at H, also the minor axis at J. The ellipse can now be drawn using the radii AH and JC. Note, however, that the ellipse is only approximate but can be used effectively on machine drawings, for example, where construction by exact geometrical methods giving a true curve could be tedious.

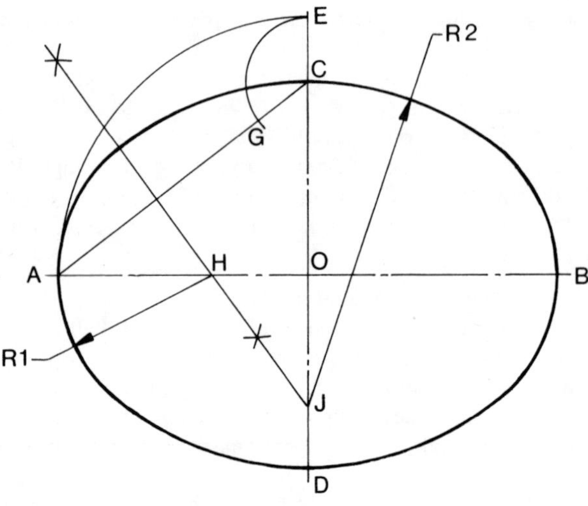

Fig. 4.6

Auxiliary views

The orthographic views which have been drawn so far have been projected horizontally and vertically but cases arise where views are required to be drawn at an angle to these axes. Typical problems occur with objects having inclined faces or features and auxiliary views are necessary to show the true shapes. A true shape can be obtained by viewing at right angles to the surface.

Consider the simple block illustrated in Fig. 4.7. A view drawn in the direction of the arrow will give the true shape of the surface shown as ABCD. The end view and the plan both show the surface ABCD, but the projected distances AD and BC are different in both views and dimensionally less than the true length.

Note that the distances AB and CD in the auxiliary view can be taken directly from the given plan or the end view as the length will be dimensionally correct in each case.

A further example is shown in Fig. 4.8 to indicate the true shape of an inclined feature. Dimensions relating to the feature would be added to this view.

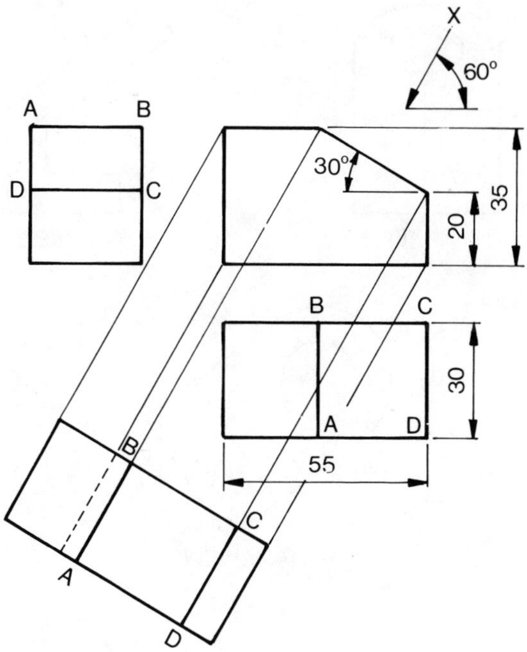

Fig. 4.7 The auxiliary view in the direction of arrow X gives the true shape of surface ABCD.

In addition to using auxiliary projections to obtain the true shape of inclined surfaces, you will also need to be able to project at angles to the vertical or horizontal. Fig. 4.9 below shows the auxiliary view obtained by projecting a view from the plan in line with the arrow parallel to the horizontal plane and at 45° to the vertical plane. Copy this solution using the dimensions given in Fig. 4.7, and note that the height shown as dimension 'Y' is taken from the front view.

The method of projecting a circle in an auxiliary view is illustrated in Fig. 4.10. Commence by drawing the centreline of the auxiliary plan at right angles to the line of sight of the arrow and in a convenient position. A typical chord P_1P_2 is shown and the distance 'X' must be obtained from

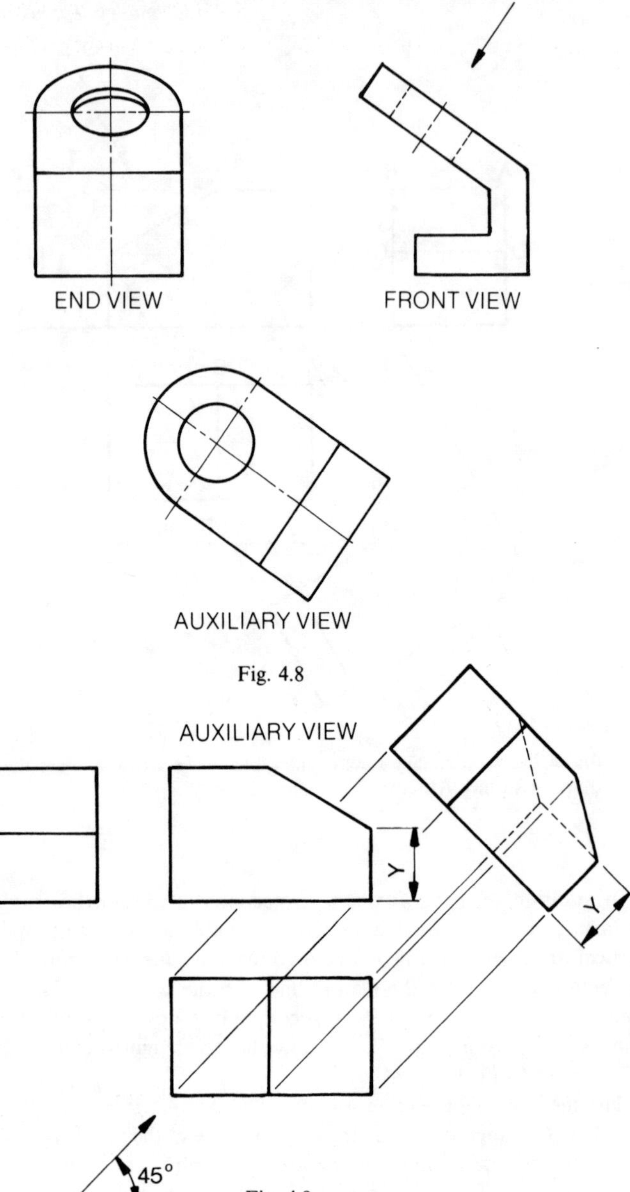

Fig. 4.8

Fig. 4.9

the true plan view using the projection lines. The curve in the auxiliary view will be an ellipse.

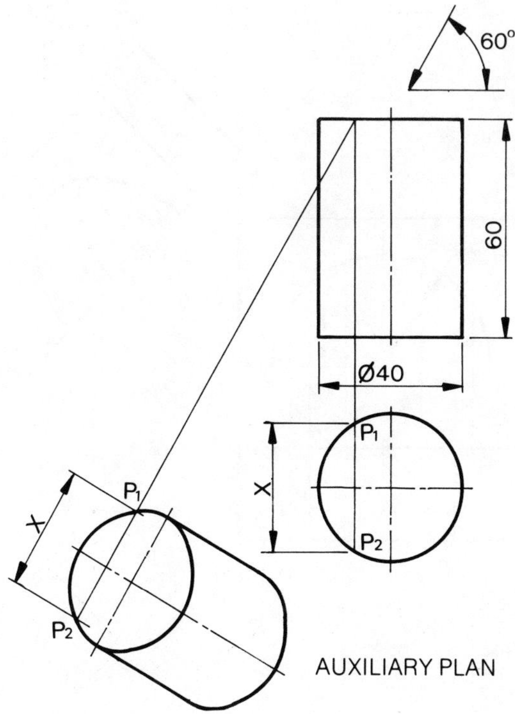

Fig. 4.10 Auxiliary projection of a solid cylinder.

An auxiliary view in the direction of arrow 'Z' is drawn in Fig. 4.11 to illustrate the method of projecting an arc. Draw the component and divide the arc into a convenient number of parts. Note that the vertical distances to the arc in the front view will be the same in the auxiliary view but projected in line with the arrow 'Z'. The complete construction has been included for the front of the arc. Distances to the curve at the rear, which can only be partly seen in the auxiliary view, are the same as those at the front and one typical point is dimensioned by the letter 'Y'.

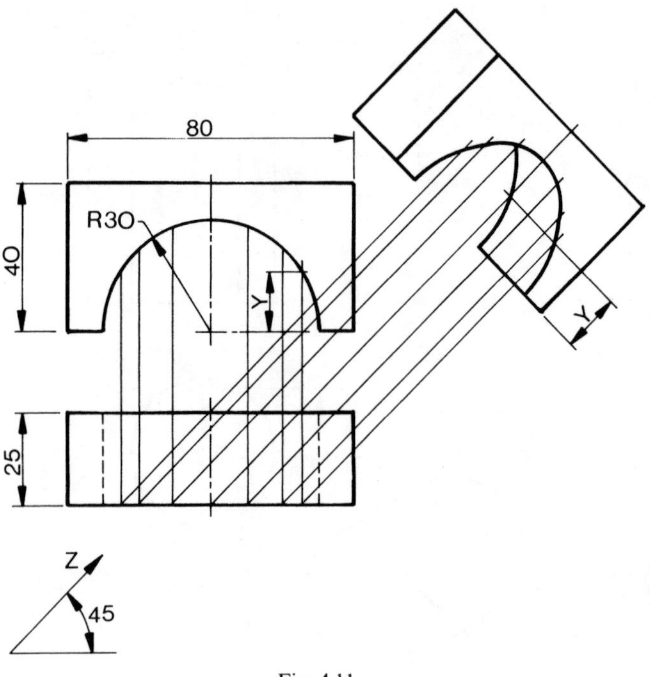

Fig. 4.11

Sections through simple solids

Examination questions are often set which test the students ability to draw sections through simple solids and the following are typical problems of a general nature. A vertical section may be assumed to be parallel with the vertical plane and Fig. 4.12 shows a possible section from a cube.

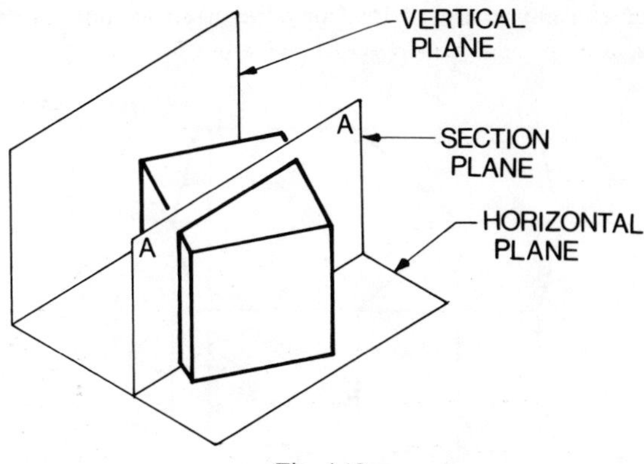

Fig. 4.12

Example 1 Fig. 4.13. Note that the cube has been cut into two parts along the section plane. The projected view contains only the part in line with the arrows 'AA'.

The actual cut surface is cross hatched by 45° lines at equal pitch. Copy the solution, as shown.

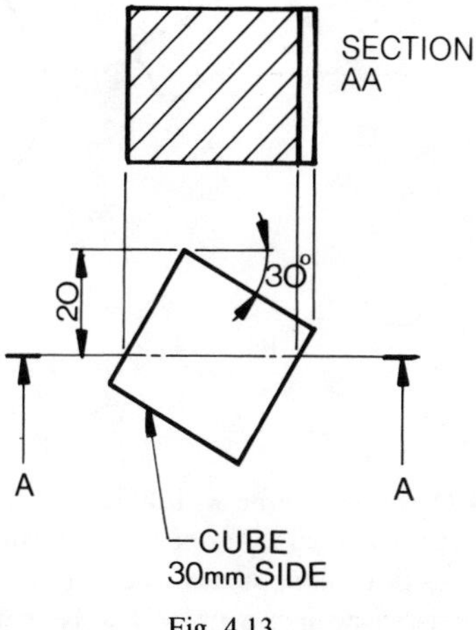

Fig. 4.13

Example 2 Fig. 4.14 shows a hexagonal prism resting on the horizontal plane with its axis inclined at an angle to the vertical plane. The arrangement is shown pictorially in Fig. 4.15. Copy the given sectional elevation and plan view.

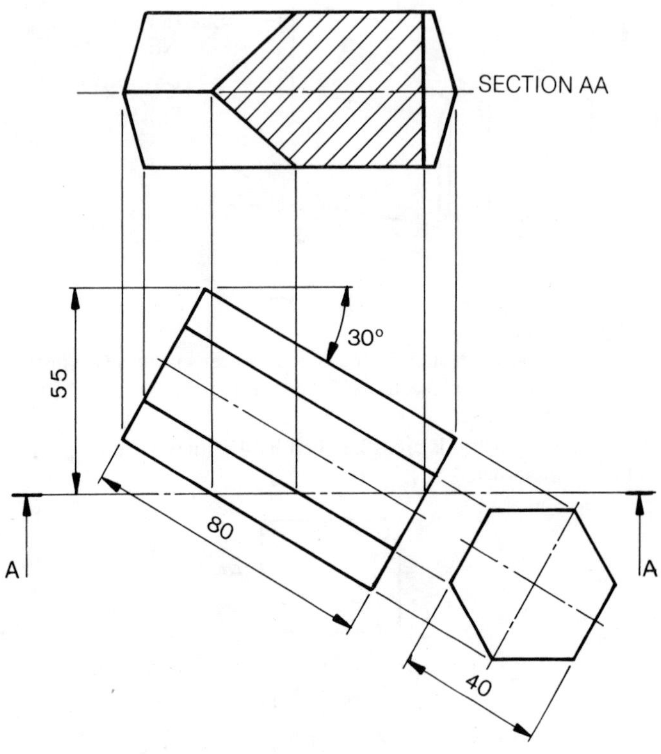

Fig. 4.14

Example 3 Fig. 4.16 shows a cube with a 30 mm side which rests on the horizontal plane, but turned, so that one side is positioned at 60° with the vertical plane. The section plane is also inclined at 30° to the horizontal plane. The problem is to project a plan view beneath the given view, of the part of the cube that remains after being cut along the section plane 'AA'. In addition, a sectional end view is required at the right hand side of the given view.

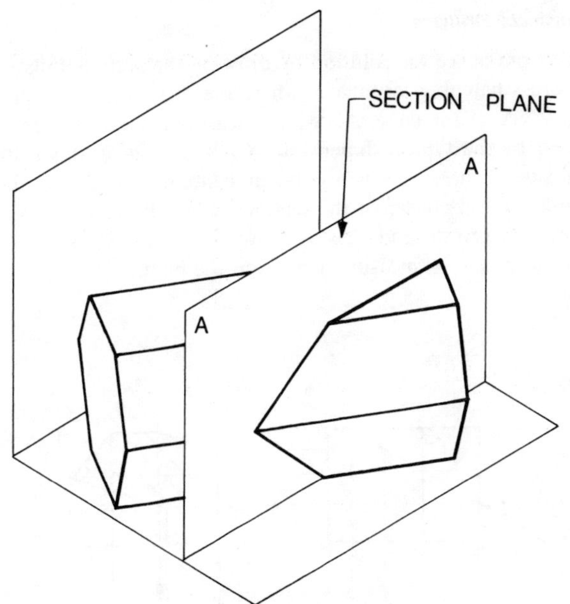

Fig. 4.15 Pictorial view of hexagonal prism.

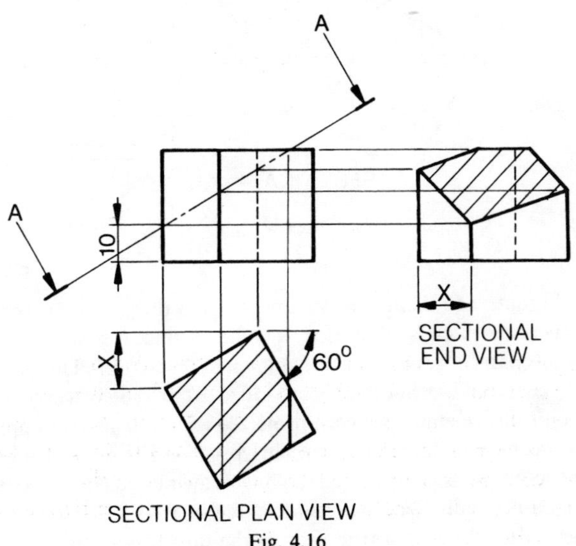

Fig. 4.16

Solution Commence the solution by drawing the sectional plan and note that the cross hatching terminates where the section plane intersects with the top surface of the cube. All width measurements in the end view are represented by the typical dimension 'X'. In preparing this solution, the left hand side of the end view was taken as a datum and this edge corresponds with the top corner of the sectional plan. Project lines from along the section plane to the end view as shown, and fix the position of the five corners in this view by measurement from the plan.

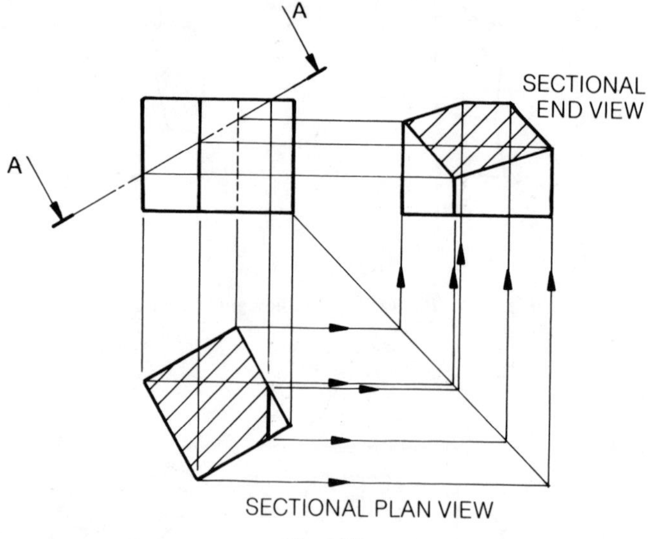

Fig. 4.17

As an alternative to taking measurements from the plan view, the following construction, which is often used, is included since it is helpful to understand the mechanical process of drawing end views from plan views. Note that a 45° construction line has been added from the corner of the cube and the points in the plan have been projected round to give the appropriate edges and corners in the end view. This practice is time consuming and cannot be used with complicated layout drawings and the experienced draughtsman normally measures as indicated in Fig. 4.16.

The method of drawing a true view in the direction of the arrows 'AA' on the cutting plane is shown in Fig. 4.18. The position of the corners of the cube in relation to a datum line XY are given by typical dimensions B and C. To draw the section 'AA', position a new datum X_1Y_1 in a convenient place and project lines from each of the corners in the front view. The datums XY and X_1Y_1 can be understood to represent planes which are both perpendicular to the horizontal plane and the dimensions B and C in the true plan and the section 'AA' will be equal in length.

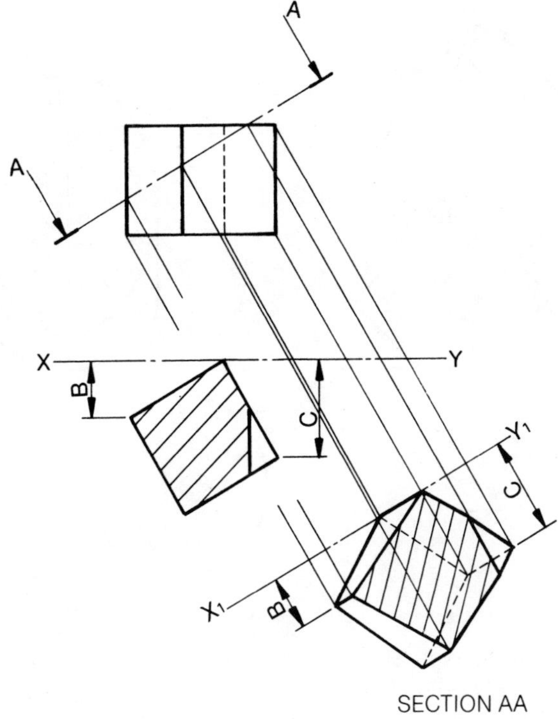

Fig. 4.18

Fig. 4.19 shows the method of drawing a section through a cylindrical solid which is inclined to the horizontal plane but whose axis is parallel with the vertical plane. Set out the cylinder as illustrated and add a circle which represents the true plan viewed in the direction of the axis of the cylinder. Let P be a typical point at any position along the cutting plane. The width across the section in the end view can be obtained from the distance across the chord in the plan view and this is given by the dimension shown as 'D'. Take a succession of points along the cutting plane and repeat this construction to give the required section.

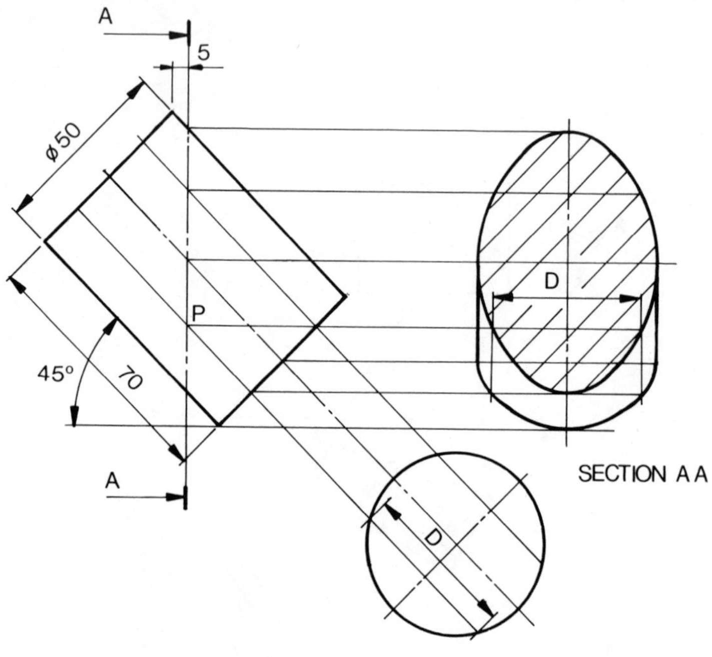

Fig. 4.19

Example 5 F 4.20 shows a ring of circular cross section, such as a doughnut and it is required to draw a plan view taken as a section on the cutting plane 'AA'. Typical construction lines are given. Let P and P_1 be any points along the cutting plane. Draw radial lines from the centre of the ring through these points and insert circles which represent the radial cross section. At P and P_1 draw chords at right angles to the radial lines and the lengths of the chords, given as dimensions C and C_1 should then be transferred to the plan view. Repeat this procedure for a number of points to give the curve illustrated. Cross hatching has been omitted for clarity.

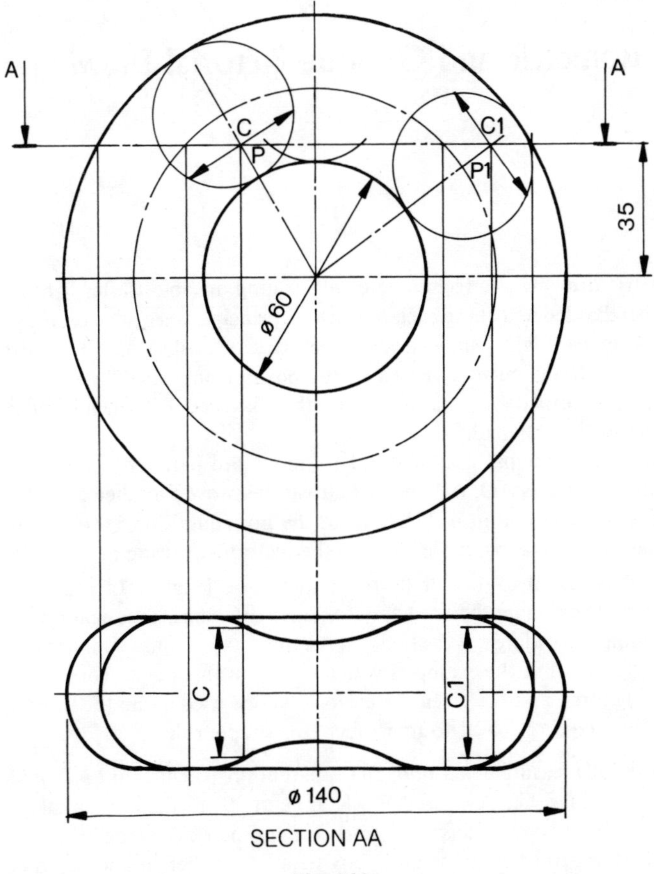

Fig. 4.20

5

Isometric and Oblique Pictorial Drawing

Isometric drawing is a technique of illustrating an object after it has been tilted so that lengths along each of its three principal axes have been equally foreshortened. Fig. 5.1 shows a very simple case of a cube with a hole through the centre. If the cube is viewed in the position indicated by the corners PQRS and turned by rotating about the diagonal PR, then a pictorial illustration can be made.

The three principal axes referred to above are parallel to the edges of the cube shown as AO, AP and AR. It can be proved mathematically that if the angle of rotation is 35°16′ from the horizontal, or 54°44′ from the vertical, then the apparent lengths on the pictorial drawing of AO, AP and AR will be equal to $\sqrt{\frac{2}{3}}$ or 0.816 of their true lengths. In practice any convenient scale may be used for isometric drawings, thus ignoring foreshortening, providing it is the same on the three axes. Isometric means 'equal measure'. In all of the examples which follow in this book, isometric scale will be ignored and the given dimensions will be used along the three axes.

At this stage you need to remember two simple rules:

1 Make all measurements along the apparent vertical lines like AO or along the two lines like AP and AR which lie at 30° to the horizontal.
2 It is often advantageous to commence an isometric drawing by imagining the object to be contained in a box with length, height and depth equal to the maximum dimensions of the object.

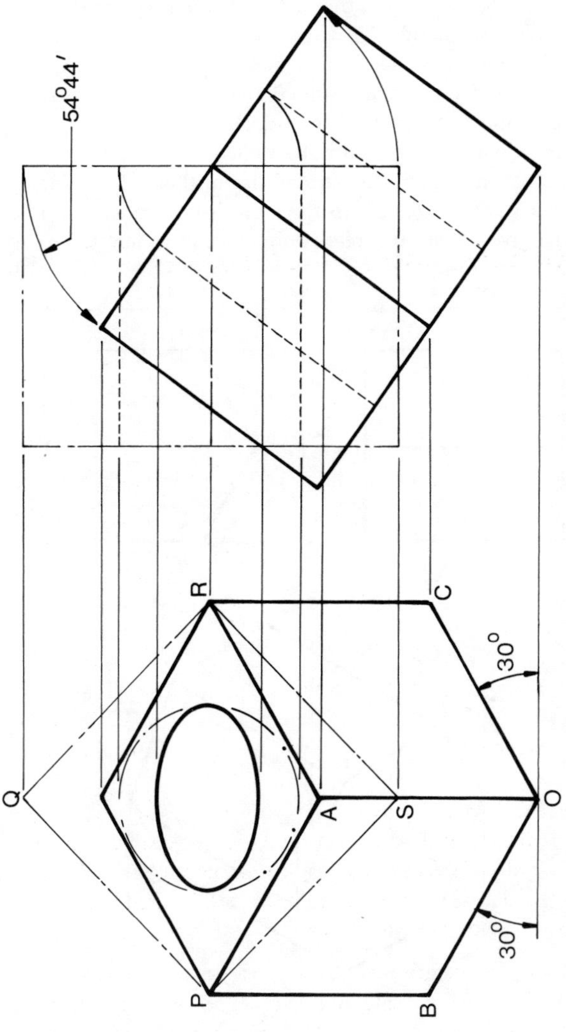

Fig. 5.1 The left hand pictorial illustration of the cube is a typical example of an isometric drawing.

Draw the examples which follow and use the two simple rules.

Fig. 5.2 shows a length of bar with a hexagonal cross section. Make an isometric drawing by constructing a box where the width is equal to the distance across corners of the hexagon (shown here as A/C) and note that the hexagon can be constructed using a 60° set square to draw the flat sides around the circumference of a 50 mm diameter circle. Insert the centrelines on one end of the isometric box and transfer the width of one of the flats, given as AB. Complete the pictorial view as illustrated below.

The example in Fig. 5.3 illustrates the method of producing sloping surfaces which lie at an angle to the main isometric axes. Commence by drawing the given front and plan views and an isometric box to enclose

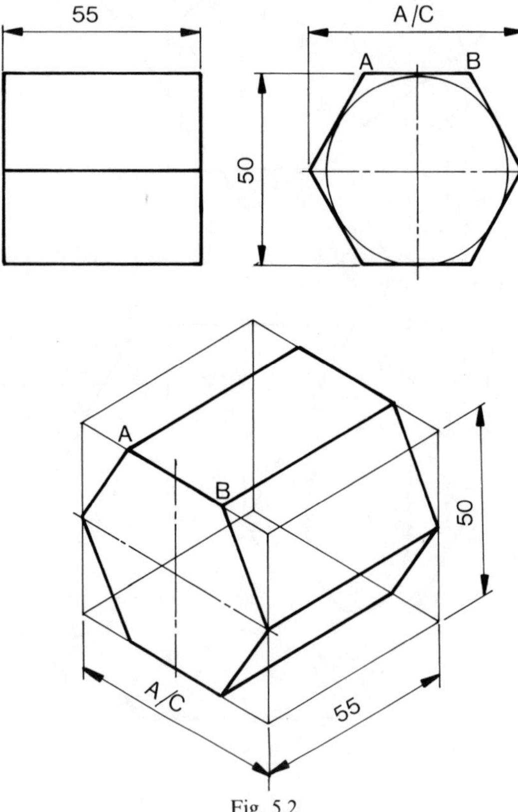

Fig. 5.2

the wedge which will scale 80 × 50 × 40 mm. The rectangle at the end of the wedge can be added by using the given dimensions measured along the axis at 30° and the vertical. The rectangle is lettered 'E'. The position of the edge shown as 'A' in the front view can now be added using dimension 'B' measured along the other 30° axis. The position of corners 'C' and 'D' should be taken from the given plan view and added to the isometric drawing.

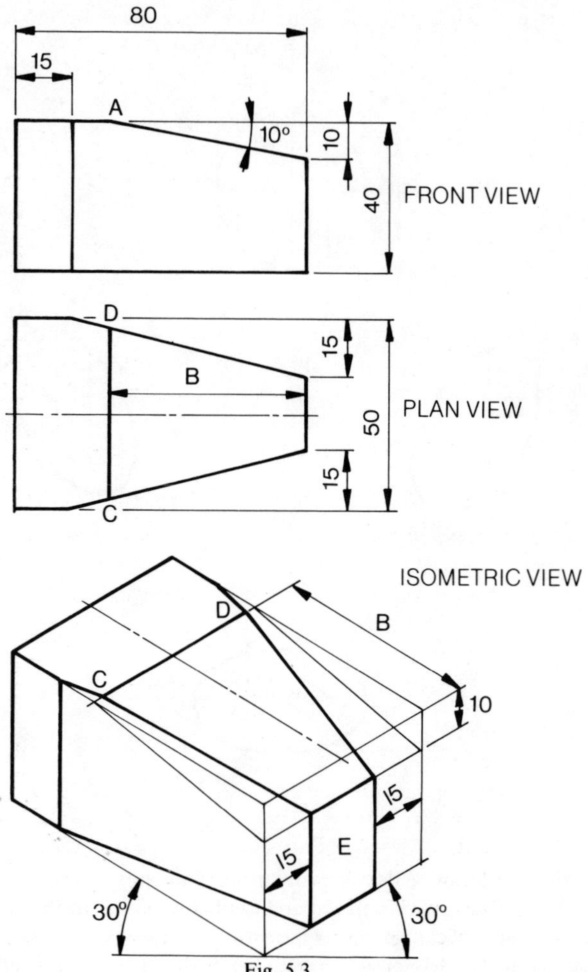

Fig. 5.3

Circles on isometric drawings

Isometric circles can be drawn accurately by the method shown in Fig. 5.4. Draw a circle, for example, 40 mm diameter and divide into 5 mm wide strips. Set out the isometric enclosing squares on the three principal axes and divide them into the same number of strips. Transfer the heights of the separate ordinates, a typical dimension being given by the letter 'A' and draw the required curves through the plotted points.

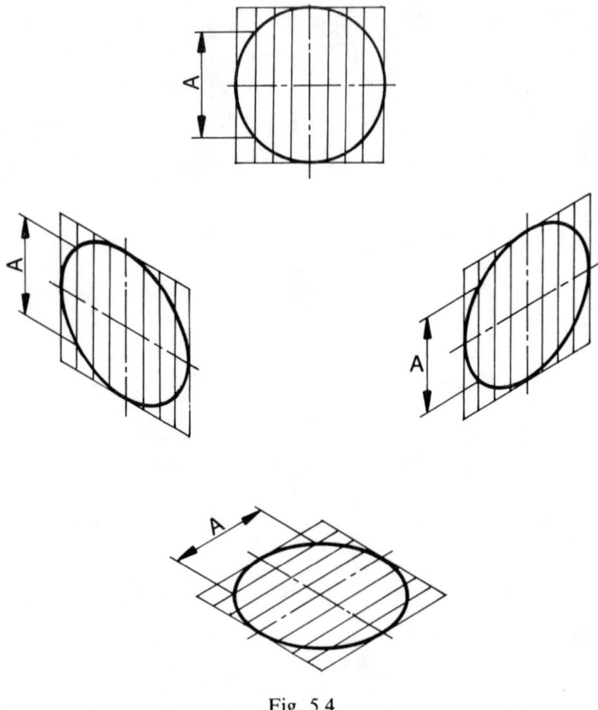

Fig. 5.4

Isometric circles can be drawn easily using templates which only require the draughtsman to position the centrelines before lining in the curves.

There are many ellipse templates available in a variety of ratios between the major and minor diameters. The ellipse template which is manufactured for use on isometric drawings will be marked 'isometric ellipses' or engraved with the angle of 35°16′, mentioned in the introduction.

The method is now shown applied to an isometric drawing of a hollow cylinder. The end elevation has been divided into strips and the heights of the ordinates transferred to the pictorial projection. Note that no hidden detail is included in the pictorial drawing.

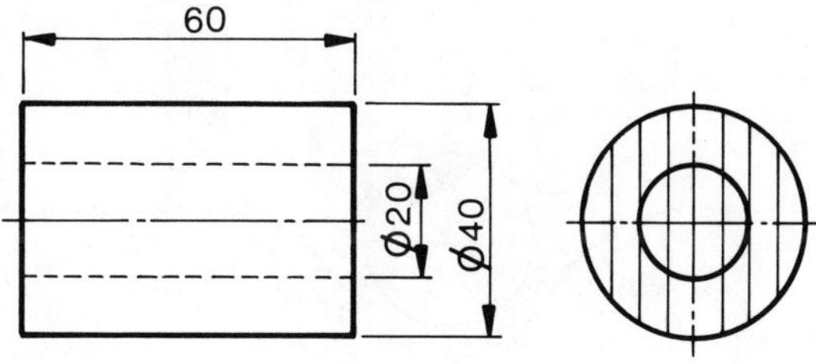

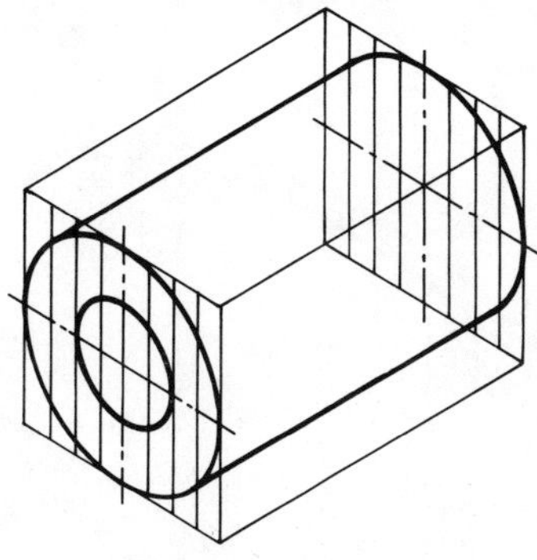

Fig. 5.5

Any curve can be drawn in isometric by the use of ordinates from a datum level. Fig. 5.6 below shows a typical profile of a lamination, 5 mm thick which is cut from aluminium sheet. Set out the profile and establish the centres for the arcs and the points of tangency. As a convenient datum, draw a line across the profile at the top edge and divide this into a number of strips.

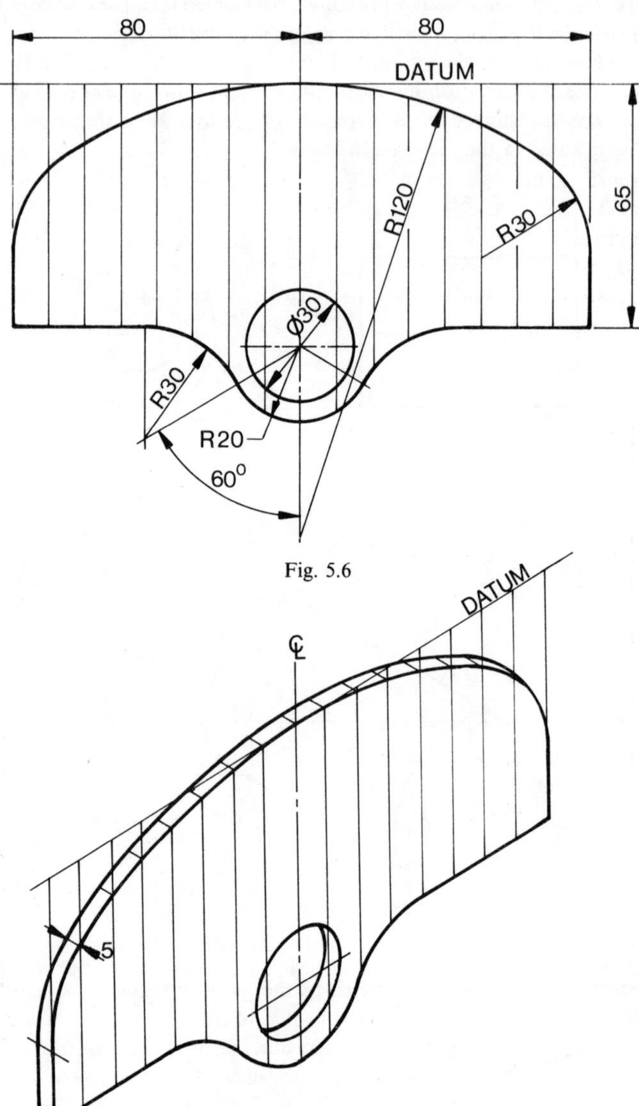

Fig. 5.6

Fig. 5.7

Transfer the height of the ordinates, measured from the datum onto the isometric drawing, once you have determined the position of the isometric axes. The solution is shown in Fig. 5.7.

As an exercise in this technique, draw an alternative solution where the datum line slopes at 30° down to the right hand side of the sheet.

Copy the worked example shown in Fig. 5.8. This problem is given to demonstrate the case where a curve is present along a surface which is not parallel to one of the principle isometric axes. The example shows a cylindrical bar which has been cut at an angle of 45°. A view square with the 45° face would give an ellipse and in this isometric drawing, is obtained by constructing vertical ordinates from points around the base. The base is constructed by drawing ordinates at 10 mm pitch, as indicated, and the

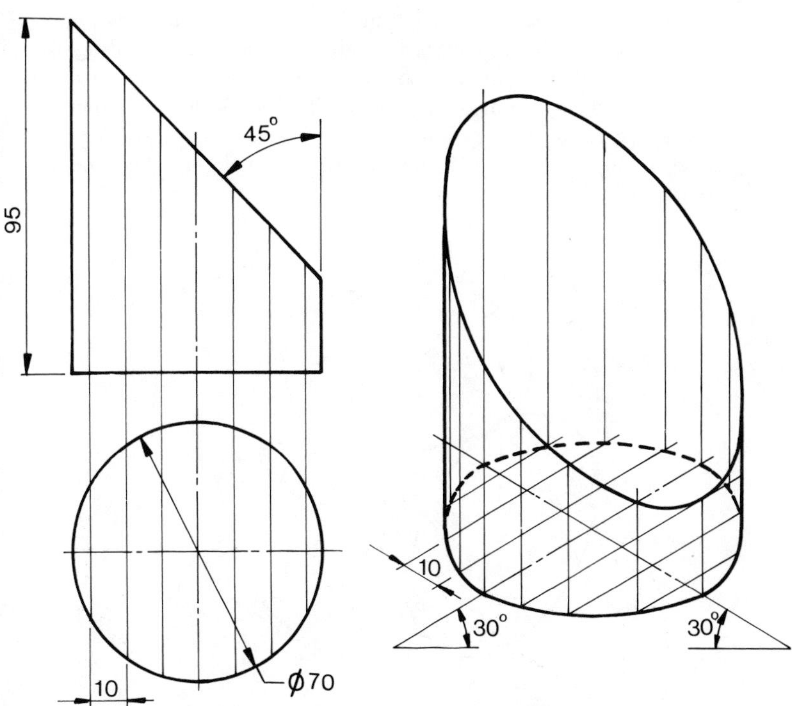

Fig. 5.8

width of the ordinates taken from the true plan view. The heights of the vertical ordinates are taken from the given front view.

Always remember that in isometric drawings, you can only measure along the three principle isometric axes and that measurements in any other direction will give distorted incorrect results.

Approximate construction for circles on isometric drawings

A simple method of constructing isometric circles is illustrated in Fig. 5.9. The circles on each of the three faces of the cube are formed by four arcs of two different radii. The construction lines giving the centrepoints for the two smaller radii can be found using a 30° set square to join the longest diagonal, which intersects with lines from the other two corners to the mid points of the opposite sides. Note the position of the tangency points between the four arcs. This construction gives an ellipse which is slightly smaller along the major axis than the true size. A useful exercise would be to draw an ellipse by the method shown in Fig. 5.4 and then superimpose this construction to compare the results. A circle size of 100 mm is suggested. This construction can be used for corner radii after fixing the appropriate centrepoints for the arcs.

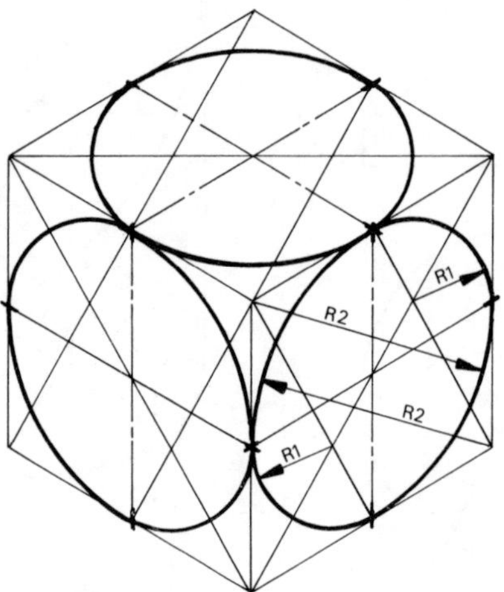

Fig. 5.9 Tangency points are shown by the thick dashes across the outlines and these are added to indicate where each of the four arcs join.

Representation of screw threads on pictorial drawings

Screw heads can be represented on pictorial drawings by a series of ellipses or circles, spaced along the axes of the threads. Ellipses will appear on isometric drawings and circles on some oblique views. The ellipses or circles should be evenly spaced to give a good artistic impression although it is not necessary to reproduce the exact thread pitch.

Fig. 5.10 shows an internal and external thread application. Make an isometric drawing of the bracket and pin using the following dimensions.

Pin

Overall length of pin	75 mm
Head thickness	12 mm
Head diameter	40 mm
Thread length	20 mm
Thread diameter	24 mm

Bracket

Bracket length	60 mm
Bracket height	60 mm
Bracket width	40 mm
Distance to tapped holes from ends of bracket	20 mm
Bracket thickness	6 mm

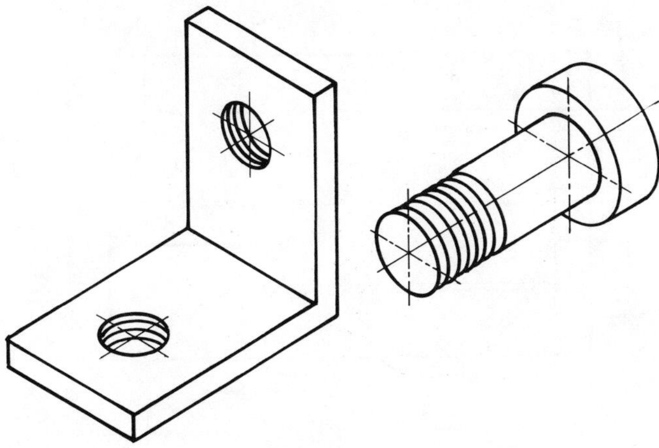

Fig. 5.10

The following questions are reprinted by kind permission of the Examination Boards and represent typical isometric drawing problems. Unless it is specifically stated, the orientation of the component is left to the student's discretion. Solutions are given for each of the problems but as an exercise draw each of the articles with a different corner in the foreground.

1 The drawing shows a plan and elevation of a photo-frame constructed in wood. Draw the frame, full size, in isometric projection. Make the point marked 'C' the lowest point of your drawing. Do not draw the glass. Do not show hidden detail.

Middlesex Regional Examining Board

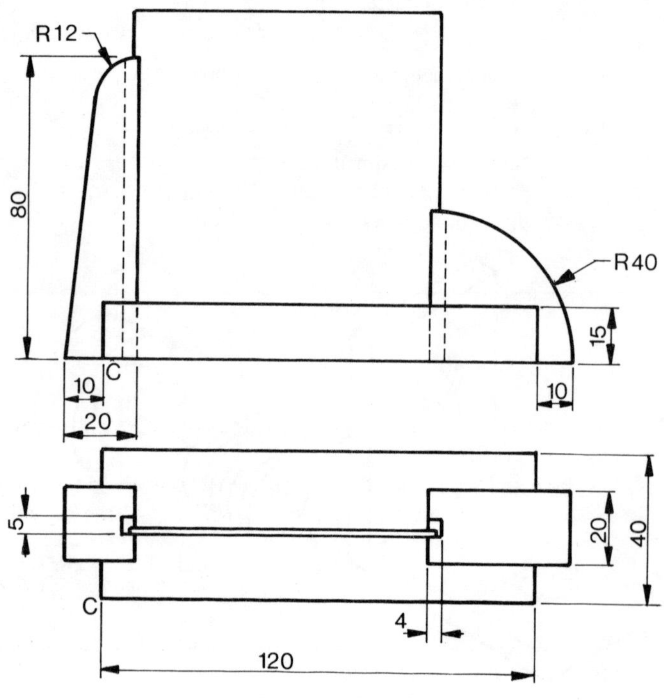

Fig. 5.11

2 The front view and plan in Fig. 5.12 show details of a Drill Table Retaining Ring. Make a full size isometric drawing of the ring.

Middlesex Regional Examining Board

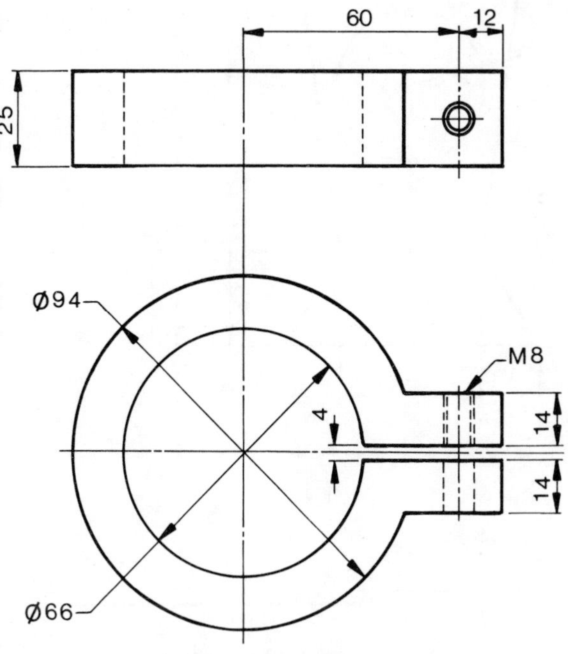

Fig. 5.12

3 Two views in first angle orthographic projection of a Presentation Shield are shown in Fig. 5.13.

Do not copy the given views, but draw full size, an isometric projection of the Shield, indicating clearly the position and size of the medallion. No hidden detail is required.

East Anglian Examinations Board

4 Fig. 5.14 shows two views of a Turn Key. Make a full size isometric drawing of the Key.

Middlesex Regional Examining Board

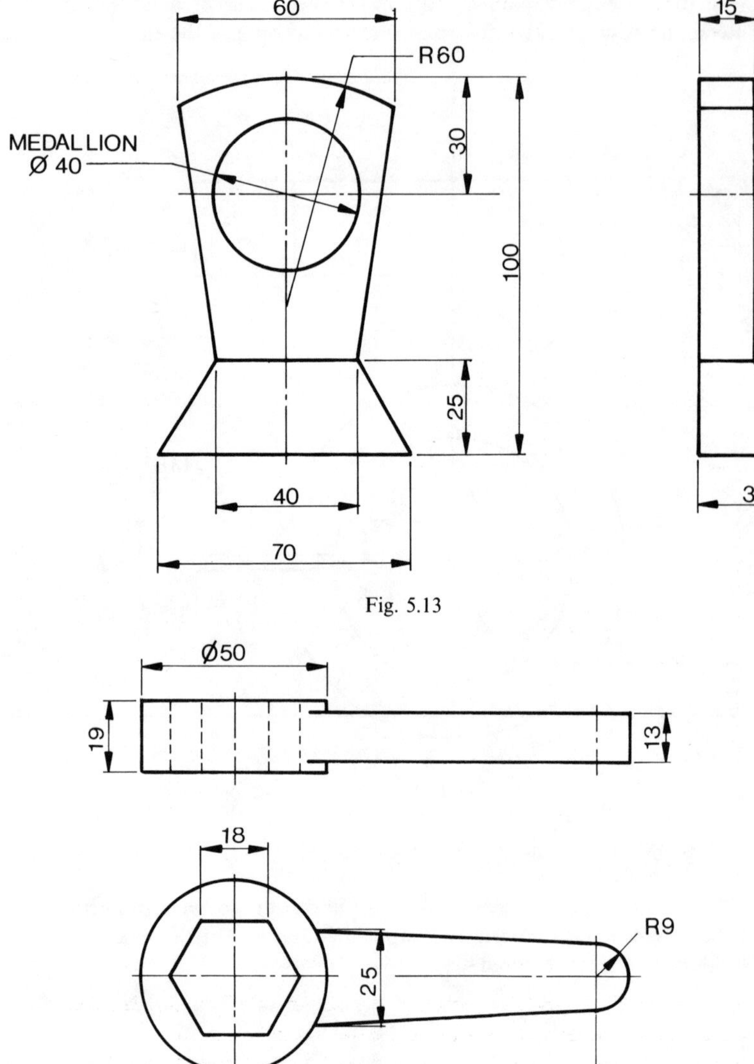

Fig. 5.13

Fig. 5.14

5 From the given orthographic projection, make a full size isometric drawing of the Ink Stand shown in Fig. 5.15.

Middlesex Regional Examining Board

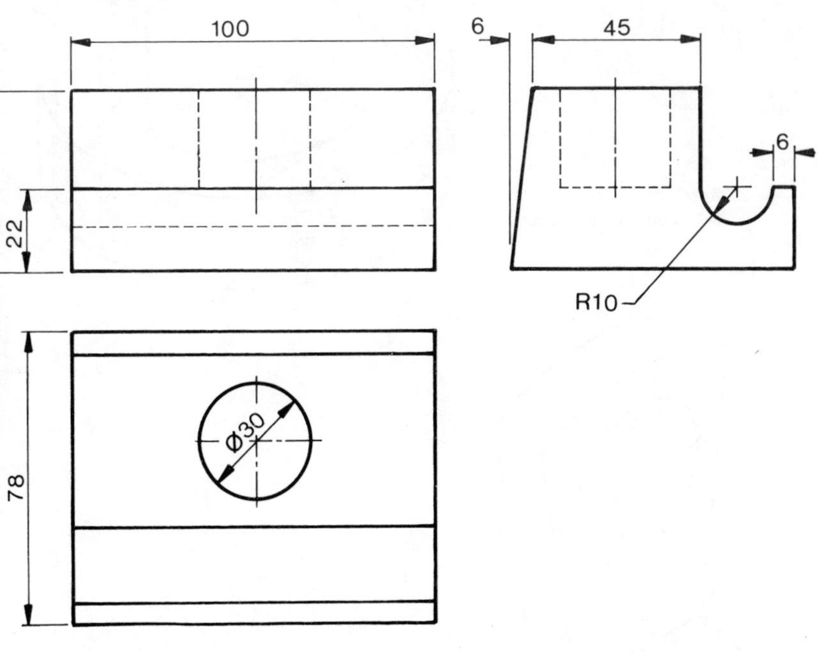

Fig. 5.15

6 Two views in first angle orthographic projection of a Support Bracket for the poles of a caravan bunk bed are shown in Fig. 5.16.

Draw, full size, to the dimensions given, an isometric view of the bracket showing all constructions clearly. No hidden detail is required.

East Anglian Examinations Board

Solutions to isometric problems

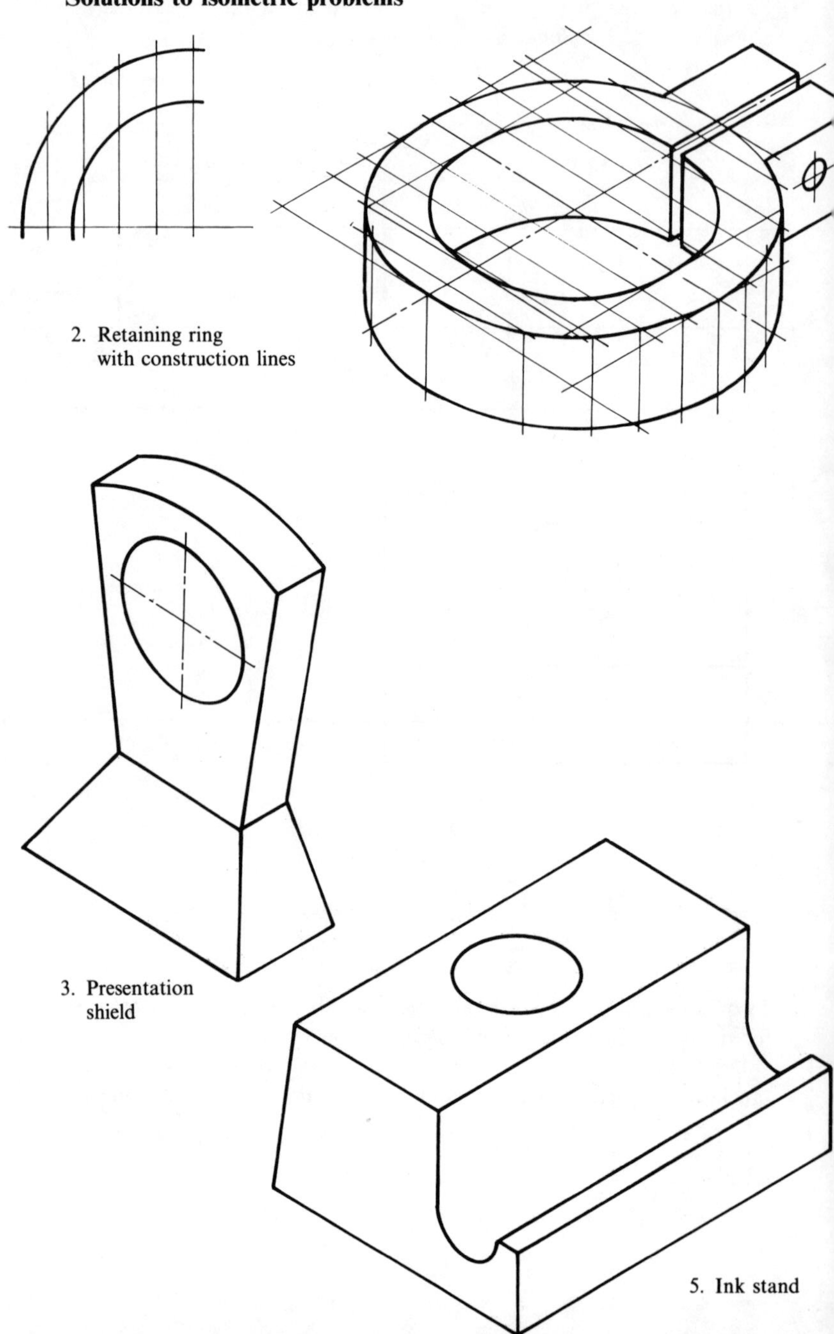

2. Retaining ring with construction lines

3. Presentation shield

5. Ink stand

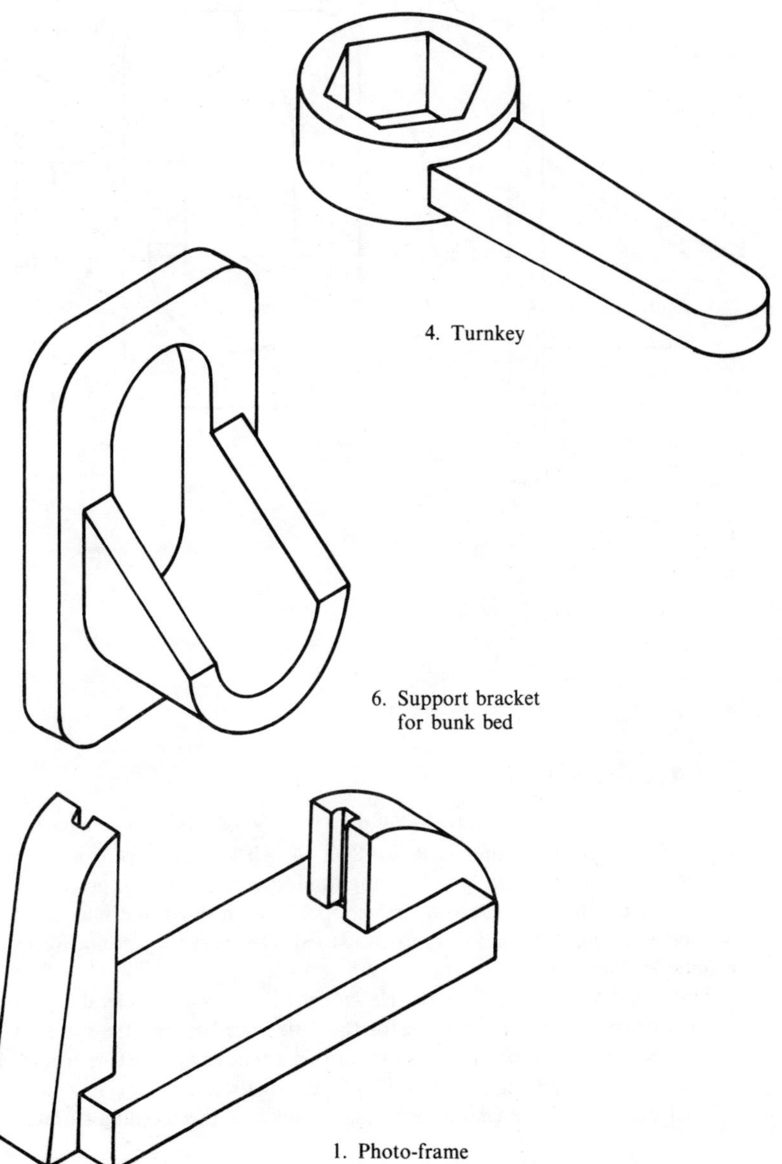

4. Turnkey

6. Support bracket for bunk bed

1. Photo-frame

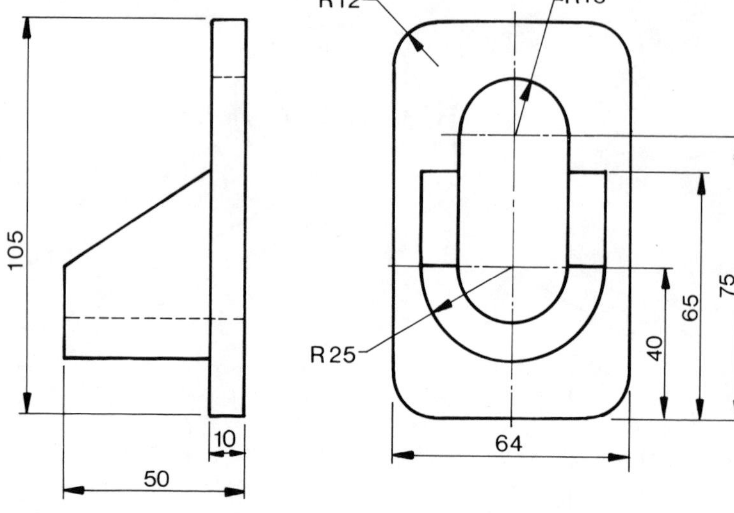

Fig. 5.16

Oblique drawings

Oblique projection is probably the simplest method of producing a pictorial drawing since surfaces directly in front of the observer will be similar in appearance to the front view in an orthographic presentation. The projectors which give depth to the drawing are parallel to each other and may be at any angle but the angle of 45° is generally used. There are two specific forms of oblique drawings.

Fig. 5.17 shows a cube with a circle on each face drawn in Cavalier and Cabinet projection and the reason for the difference between them is that since there appears to be distortion by drawing the rear projectors true to size in the Cavalier method, then the lengths of these lines are halved in Cabinet drawings. Note that in both cases circles on the receding surfaces appear as ellipses and it is therefore advantageous to orient an object so that circular features appear in the frontal plane, if possible.

Ellipses can be drawn using templates or by the constructions which follow. (See figs. 5.18 and 5.19, p. 136.)

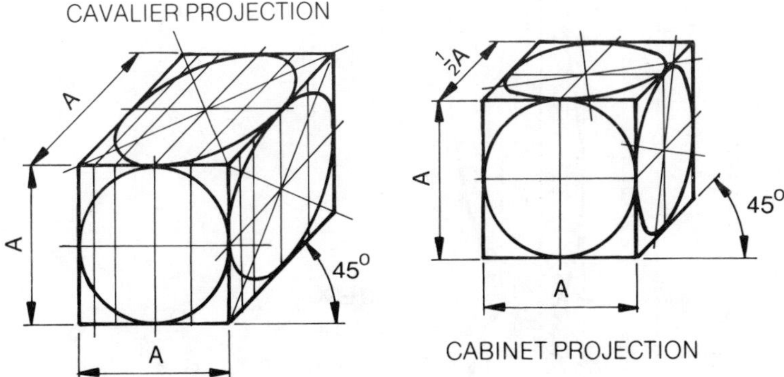

Fig. 5.17

As an exercise in Cabinet and Cavalier projection, draw two cubes and use the constructions shown below to produce ellipses on the receding faces. Take dimensions shown as 'A' in Fig. 5.17 as 100 mm.

Many examination questions do not advise the student which type of oblique drawing is required but it is the authors' experience that solutions presented as cabinet drawings at the 45° angle will always be acceptable unless instructions are given to the contrary.

Setting out an oblique drawing

It was mentioned earlier in the isometric drawing section that it is helpful to imagine the component contained in a box which was equal in size to the overall dimensions of the component. Consider the Support Bracket shown in Fig. 5.16. Quite clearly there are two separate box shaped elements in its construction and stage 1 below shows how these would be set out with measurements taken from the front face which is our datum in this case. Stage 2 involves adding the various radii bearing in mind the thickness of the material at the corners. The solution is shown in cabinet projection at 45°. Copy this solution as a good example of oblique drawing.

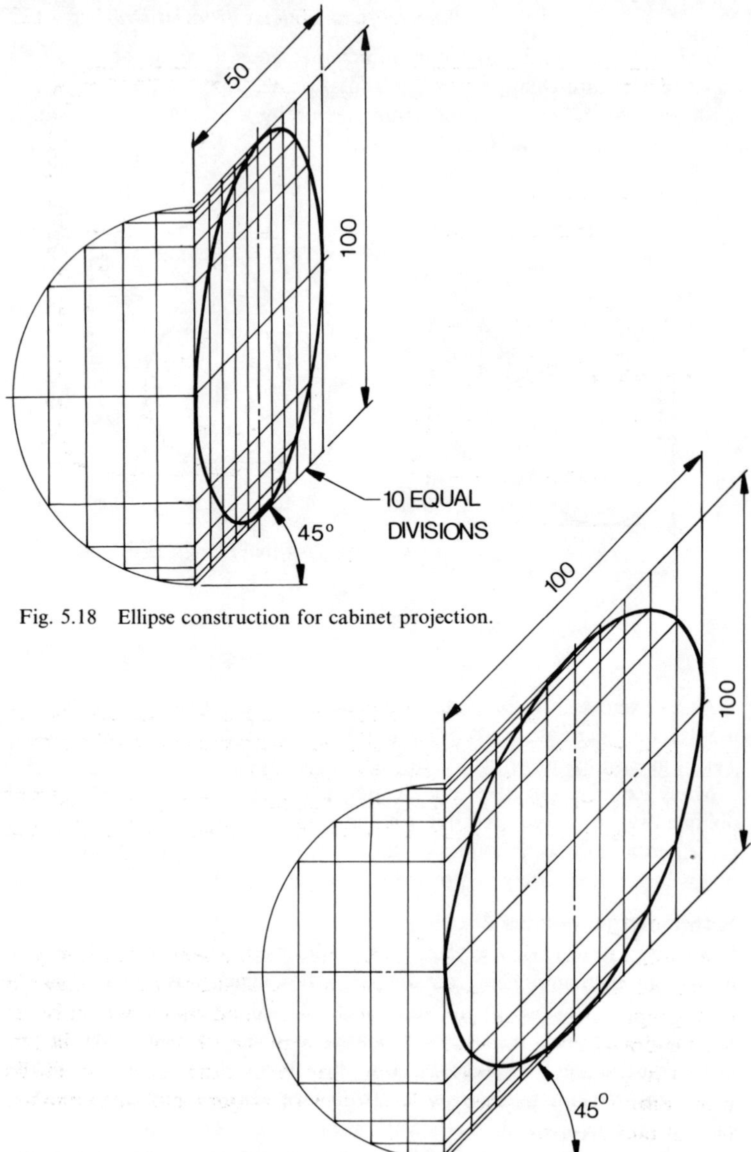

Fig. 5.18 Ellipse construction for cabinet projection.

Fig. 5.19 Ellipse construction for cavalier projection.

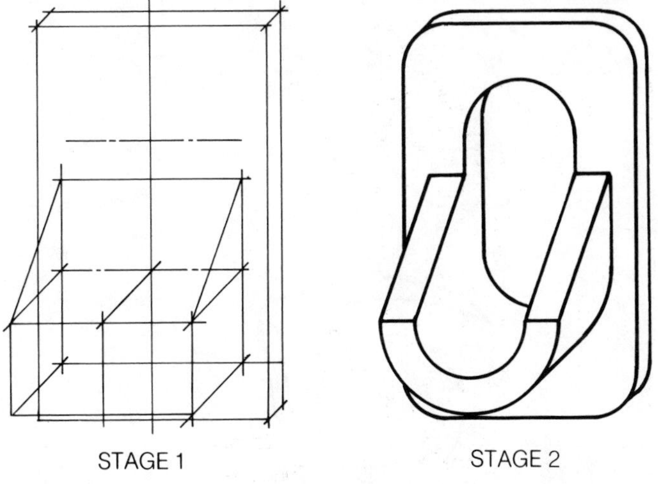

Fig. 5.20

For further experience copy the illustrations below which are oblique cabinet drawings of the turnkey, shield and photo-frame using the dimensions given previously in Figs 5.14, 5.13 and 5.11.

Draw also oblique solutions for the components in Figs 5.15 and 5.12 and note that the arrangements below give the minimum number of plotted curves.

Sketching

The design of every product probably starts with a sketch which may be drawn freehand or with the aid of instruments. Sketches may be drawn in orthographic or pictorial projection and the method chosen would be left for the draughtsman to decide. It is important that sketches must be presented in good proportion otherwise they do not convey a true lifelike impression. Lack of proportion is probably the major factor in the production of poor sketches.

Line thicknesses for sketches should be the same as for conventional engineering drawings. Generally hidden detail is not shown. Dimensions

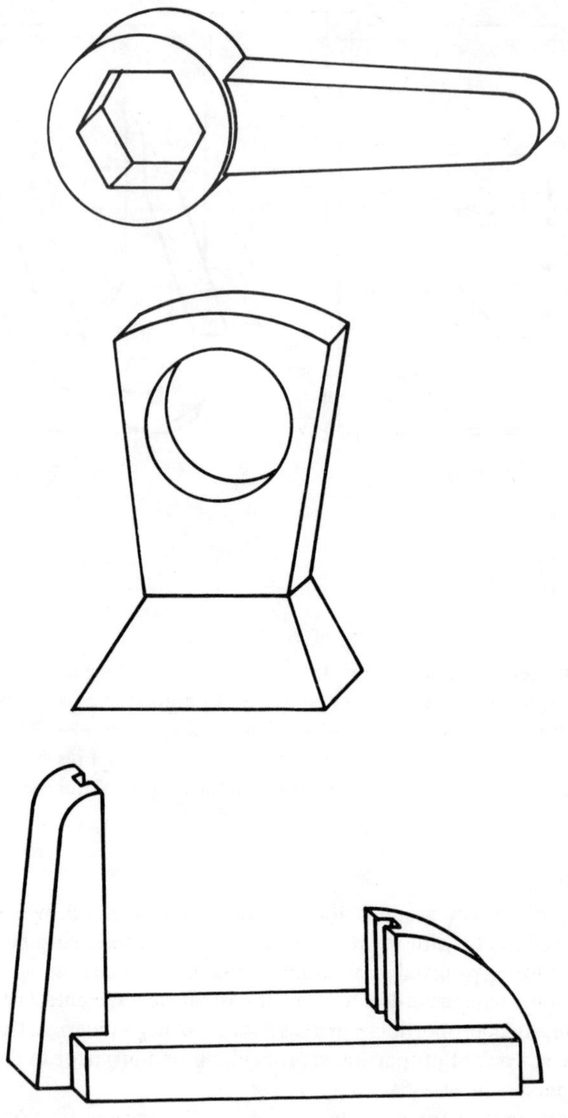

Fig. 5.21

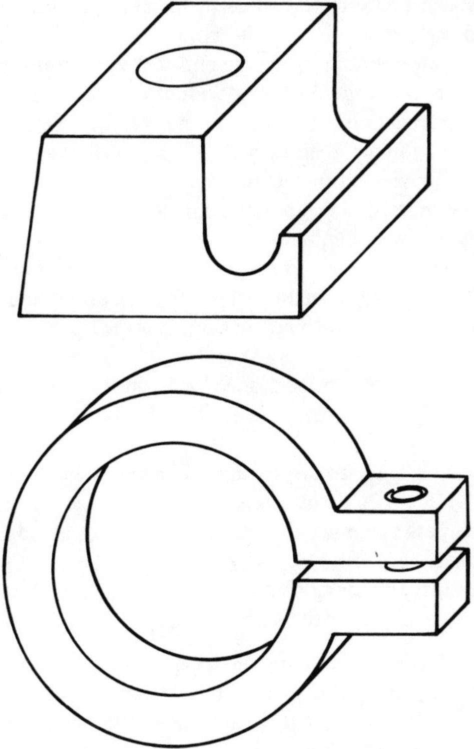

Fig. 5.22

can be added to pictorial sketches to convey numerical information but it is not advisable under any circumstances to scale such drawings for manufacturing information, where details are missing.

For pictorial sketching, it is not necessary to have a great deal of natural artistic talent. You will not be expected to produce a photographic effect but rather a clear, well proportioned presentation of geometrical shapes on a picture plane.

Practice of course makes perfect, but the following points will assist in developing a good sketching technique.

1 Use a well sharpened soft grade pencil, for example, F or HB.
2 Do not grip the pencil close to the point but hold it lightly so as to ensure reasonable freedom of movement of the hand and wrist.
3 Study the object at length before deciding the best way to sketch it, always remembering that the purpose of the sketch is to emphasise detail. If a choice can be made, then turn the object, so that the major details appear in the foreground of the sketch.
4 Note any prominent axes or lines of symmetry. If none exists then there will probably be some dominant horizontal or vertical line or lines that can be regarded as datums and the sketch can be started by positioning these essentials. Before proceeding, check that the angles and inclinations of these datums are correct and that any axes are straight by looking along the surface of the drawing sheet.
5 Having produced a well proportioned skeleton, the secondary features can then be added. It often pays to sketch this detail lightly before lining in.
6 Most technical sketches do not require to be finished by having shading and shadows added. In many cases emphasis can be made by varying linework thickness along prominent edges but try to avoid smudgy and woolly sketches.
7 Concentrate on the drawing of uniform lines, thick or thin, straight or curved and keep all joins neat and tidy.

For sketching practice, try to copy the six illustrations given as solutions to the isometric drawing exercises on p. 132, but draw each component to full size dimensions. Remember that good proportion is very important. In order to produce rapid freehand sketches, many draughtsmen commence by drawing guide lines similar to the isometric and oblique boxes, previously mentioned, which cover the overall dimensions.

6

Intersections and Developments

Intersections

Intersections arise when one solid joins or penetrates another. A typical example is illustrated in Fig. 6.1 where two cylinders of different diameters. meet at right angles. Intersection curves need to be plotted since they do not generally form part of circular arcs.

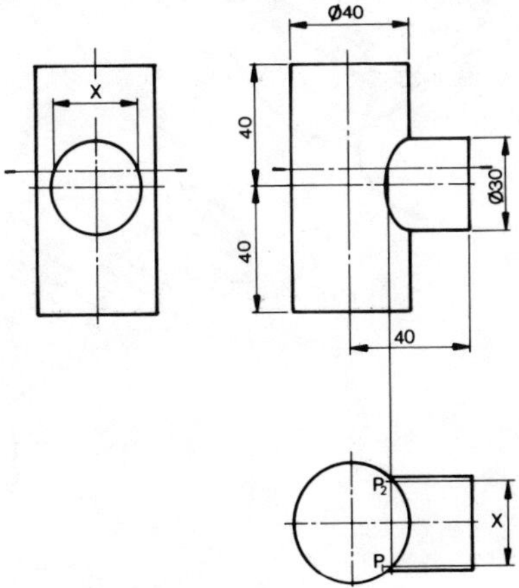

Fig. 6.1

A pictorial view of a method for finding a solution is given in Fig. 6.2. Consider taking a section, drawn at right angles to the axis of the larger cylinder and parallel to the axis of the smaller cylinder. The sectional plan will then show a circle for the vertical cylinder and a rectangle of width 'X' for the horizontal cylinder which intersect at the points marked P_1 and P_2. From the plan view, these points are projected back to the section plane in the front elevation. If this procedure is repeated with further section planes at different levels, then the curve of intersection, or curve of interpenetration as it is sometimes called, can be drawn through the points obtained.

The method described here, can be applied in solving many problems and it should be noted that any horizontal section, in this case, could have been taken to find points on the required curve. The draughtsman only therefore takes sufficient sections to produce a satisfactory result.

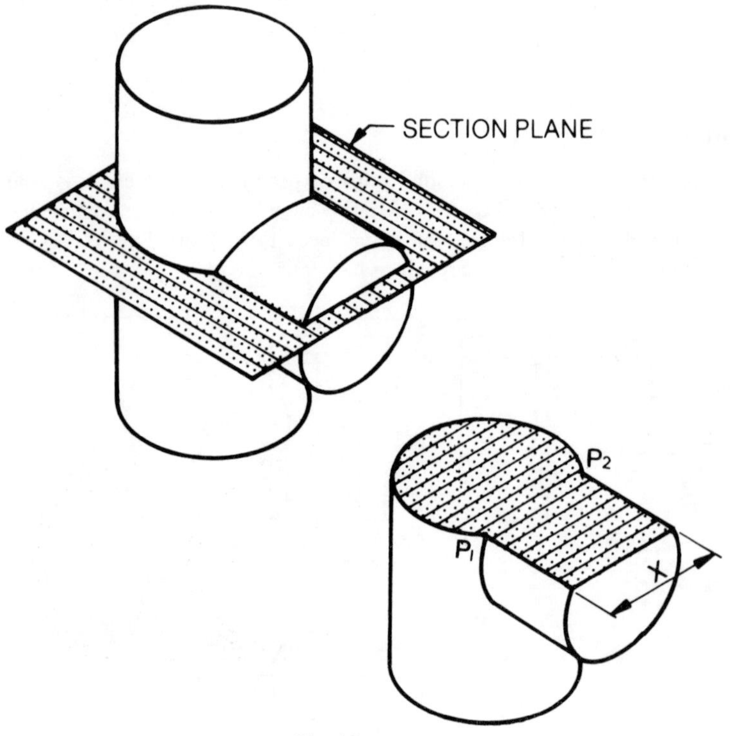

Fig. 6.2

If it was required to make a model of the tee shaped component in Fig. 6.1 from thin sheet metal, then the line of the intersection would also give the position of the join of the two parts. The method of drawing a pattern for the shapes of the two parts is given later in this section.

Intersection of hollow cylinders with similar inside and outside diameters.

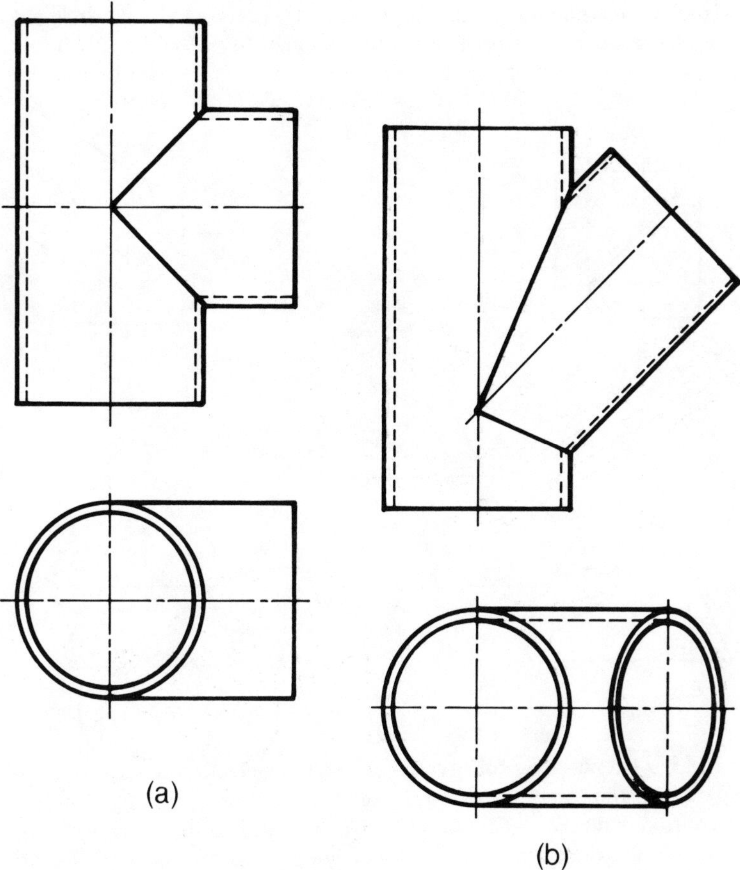

Fig. 6.3

If the cylinder and the branch have the same diameter, then the intersection will be a straight line. Fig. 6.3 shows two general cases of fittings which have been manufactured by moulding in PVC. In case (*a*), the line of intersection will be at 45° to the centrelines, in case (*b*), the line of intersection is positioned after plotting the intersections of the centrelines of the two parts.

Fig. 6.4 shows some additional cases of interest. If two cylinders, or a cone and a cylinder, or two cones intersect each other at any angle, and the curved surfaces of both solids enclose the same sphere, then the outline of the intersection in each case will be an ellipse. In the elevations shown, the join between the two parts will be a straight line.

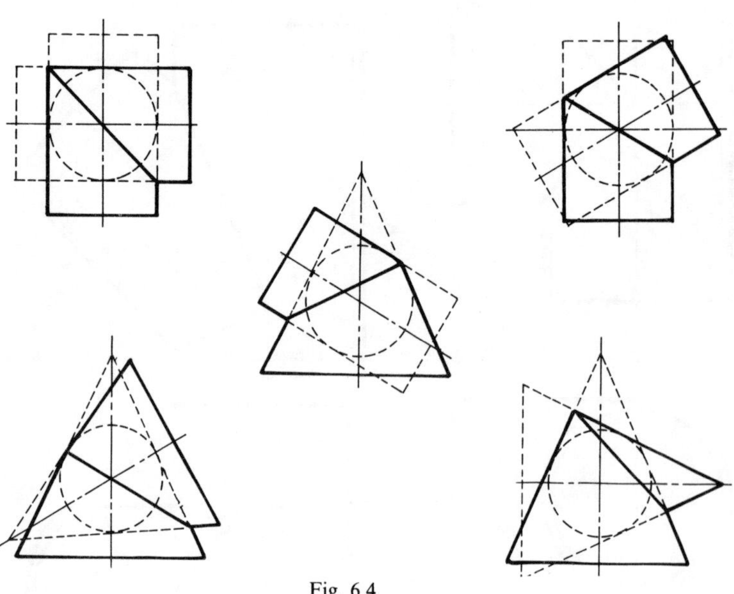

Fig. 6.4

Fig. 6.5 below shows four typical examples where an intersection forms part of the complete illustration. In each case one section plane has been added to remind the reader of the construction method for the curve. Copy these examples not only as examples of intersections but also as valuable projection exercises.

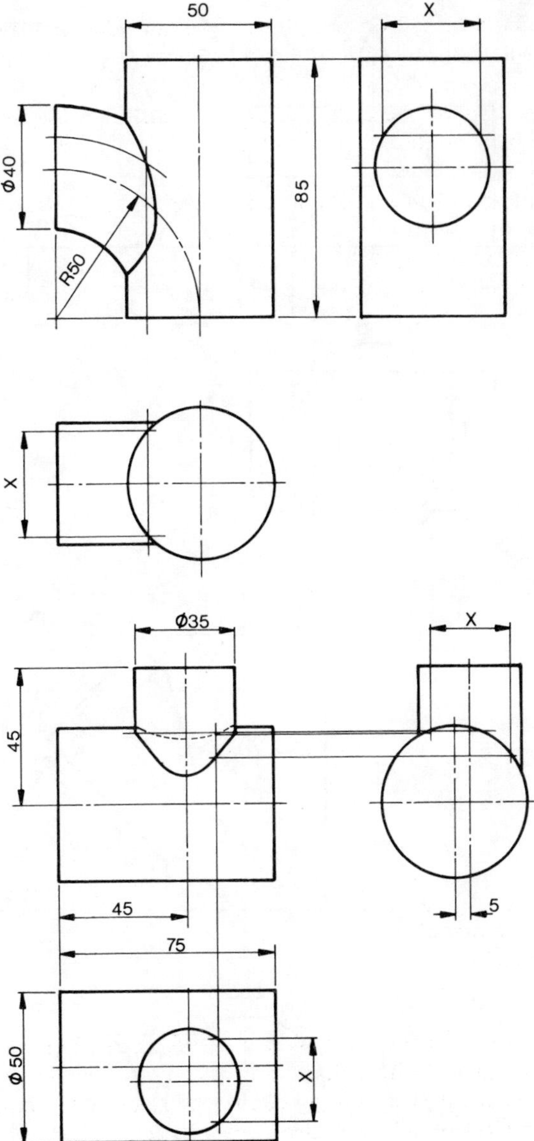

Fig. 6.5

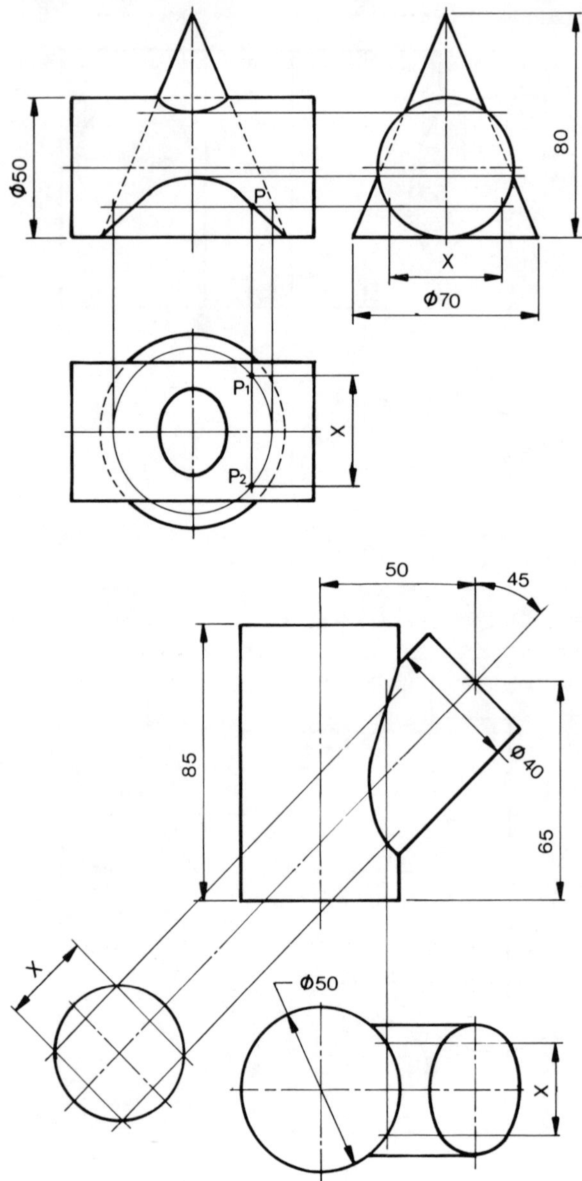
Φ50
80
P
X
Φ70
P1
P2
X
50
45
85
Φ40
65
X
Φ50
X

Curves generated by machining processes

Many components are manufactured in a machine shop and the tool which cuts the metal often leaves curves which are required to be plotted when completing a detail drawing. A typical example is given in Fig. 6.6 where a cylindrical bar of 90 mm diameter is turned in a lathe about the axis shown as AA. During the turning operation the 80 mm radius is cut and since the centrelines of the 90 mm and 50 mm diameters are not in the same line, then the result is the curve across the component and P is a typical point on the curve.

The method of drawing the curve is very similar to that described earlier. Take any section plane, and one has been indicated here as BB. Draw a circle in the end elevation of radius R and note where the circle crosses the outline of the bar at P_1 and P_2. Project a line through P_1 and P_2 back to the section line BB to give point P on the required curved line. It will be obvious from the illustration where the curve starts and finishes so it is only necessary to take sufficient other section planes, similar to BB, between the starting and finishing points to determine the route of the line. The construction does not have to be applied at regular pitches and all of the construction lines which we have used are included on this drawing. Plot the curve using the dimensions supplied.

Redraw the example in Fig. 6.7 and note the effect of decreasing the distance between axis AA and the axis of the original cylindrical bar. The machining line is constructed in exactly the same manner and all construction lines are indicated.

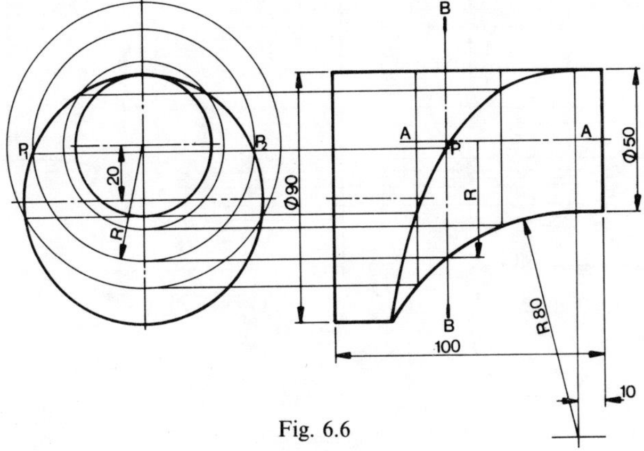

Fig. 6.6

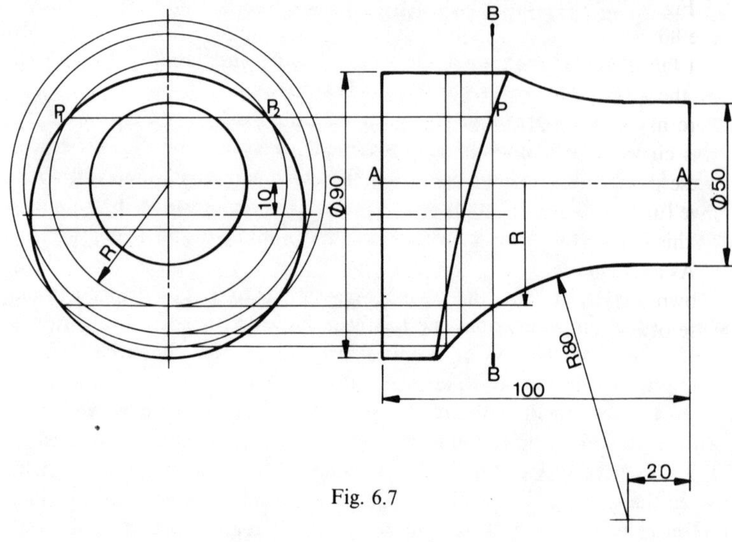
B
P1
P2
P
A
A
10
Φ90
Φ50
R
R
R80
B
100
20

Fig. 6.7

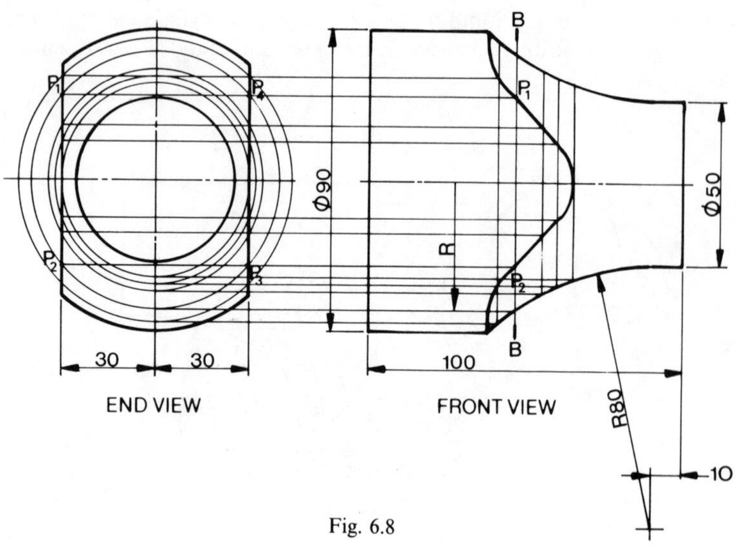
B
P1
P4
P1
Φ90
Φ50
R
P2
P3
P2
B
30
30
100
END VIEW
FRONT VIEW
R80
10

Fig. 6.8

Fig. 6.8 shows a component which has been turned on a lathe to produce the 80 mm radius and then been milled on a milling machine to leave two flat faces, at the sides, in the end view. To construct the 'S' shaped curve on the front view, it is necessary to take sections as before, and one is shown here as BB. Draw a circle in the end view of radius R, and note where this circle crosses the flat milled surfaces at P_1, P_2, P_3 and P_4. Transfer these points back to the section in the front view. Additional sections will give further points on the required curve. Curves similar to those indicated in this example can often be seen at the sides of engine connecting rods.

As a further exercise, using the same principles, redraw the part of a cone shown in Fig. 6.9 and construct the curve in the plan view. This is another case of a curve produced by a milling operation.

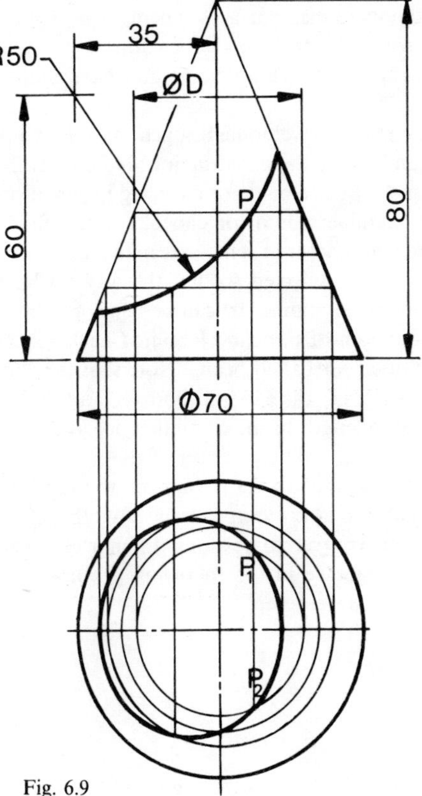

Fig. 6.9

Development

Many useful articles are manufactured from flat sheets of material. A paper pattern or development is drawn (alternatively a more substantial template may be prepared), and the required material is then cut out as a copy of the pattern and bent or formed to produce the finished product. This is a similar process to that performed by the dressmaker who cuts her fabric on a flat surface, then positions the folds and join lines before machining. In this particular aspect of draughtsmanship we are not concerned with the deformation of metal which can be performed by pressing, in the case of car bodies, metal spinning often used to produce the bell mouths of musical instruments, or deep drawing, an operation where long thin-walled containers are manufactured. In this section the development of geometrical shapes is confined only to their surfaces. There are also many different ways of joining metal so in the problems which follow we have developed the surfaces which form various components and have omitted any material used for joins.

As a simple example of development, open out the two parts of a match box, draw them and check the construction. Note carefully how the sides are kept in place and the use made of covering paper to provide the joins. Open out empty containers used for chocolates and food packaging and note how these are manufactured. There are many useful lessons to be learnt from the way shapes are selected so that the original blank material can be used in an economical manner or can be bent or formed into a specific shape with the minimum of time and labour. Check also the construction of the type of can used for baked beans. Note that the container is manufactured in three parts, i.e. the top and bottom, and the side which is one strip of metal equal in length to the circumference of the top plus the extra metal needed for the joins. The position of the join is very important on some fabricated products, often for reasons of accessibility and sometimes it is deliberately hidden to improve appearance. In other cases where a choice is possible, it is often arranged to be along the shortest edge for economical reasons and this is the case in the following problem which is a typical example of cylindrical development.

Cylindrical development

Fig. 6.10 shows part of a scale drawing of a chimney cowl which is manufactured from four separate pieces of sheet metal. If the diameters of the four circular parts are similar, then the joins will appear on the given elevation as 45° lines. The pattern for the centre section 'A' is given below.

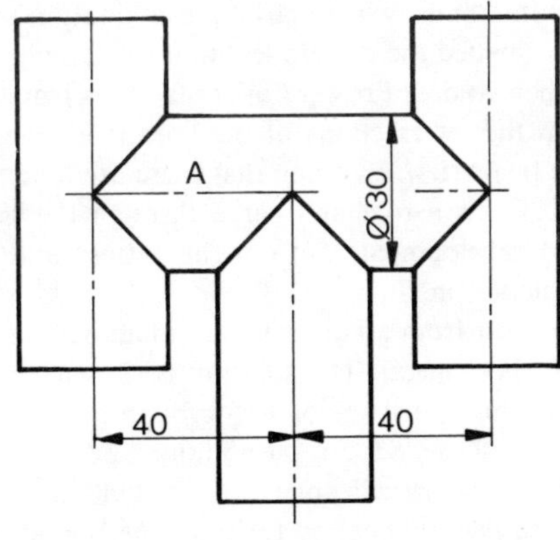

Fig. 6.10

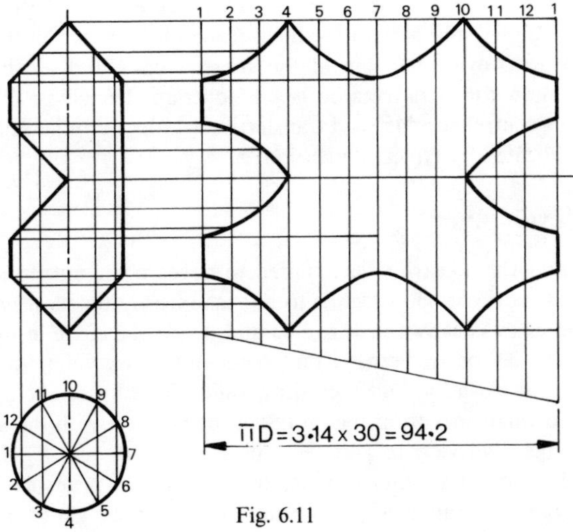

Fig. 6.11

Set out the section 'A' as shown, and add a circle in the plan position. Divide the circle into twelve equal parts, and transfer each point on the circumference up to the front view but number each point for convenience. The pattern length will be equal to the length of the circle circumference and this is given in the end view position. Divide the length into twelve equal parts, as previously described, and number each part. This is a useful practice since one often finds, when marking examination scripts, that the student has in error divided the pattern length into eleven or thirteen parts and this error can be avoided. Project horizontal lines from the front view to the pattern from the intersections of the lines from the plan with the sloping faces in the front view, ensuring that these are taken to the correct numbered position. Draw a freehand curve through the points obtained to give the required development. Cut out the pattern and bend it round to check the construction method.

Fig. 6.12 shows a vent from an air dryer at a launderette which is to be manufactured from sheet metal. The solution is shown, drawn one-tenth full size and includes the true shape of a wire mesh grill at the outlet. The shape of the grill is an ellipse. Set out the solution and note that the length of the pattern is such that it will not easily be divided into equal parts. Divide the length into twelve equal parts by the method already described and divide the end elevation circle into 30° divisions. Project the lengths of corresponding lines from the front elevation to the plan view to give the width of the pattern at each point. Draw a curve through the points obtained to complete the pattern for the outside of the vent. Note that it is customary to provide the join at the shortest edge and it would be necessary to allow for extra metal to give an overlap. The ellipse is drawn by projecting a centreline at 45° and transferring widths from the end elevation circle, indicated by typical dimension 'X'.

Conical development

Fig. 6.13 shows the construction required to develop the curved surface of part of a cylindrical cone which is to be made from thin sheet metal.

Set out the cone as shown in Fig. 6.13 and divide the circle in the plan view into twelve 30° parts. Transfer the points on the circumference to the base of the front elevation and then draw radial lines up to the apex O. Draw perpendicular lines from the points of intersection with the sloping face back to the plan view to give the typical points P_1 and P_2. Line in the ellipse through these intersections to complete the true plan view. Lay out an arc with a radius which is equal to the slant height of the cone

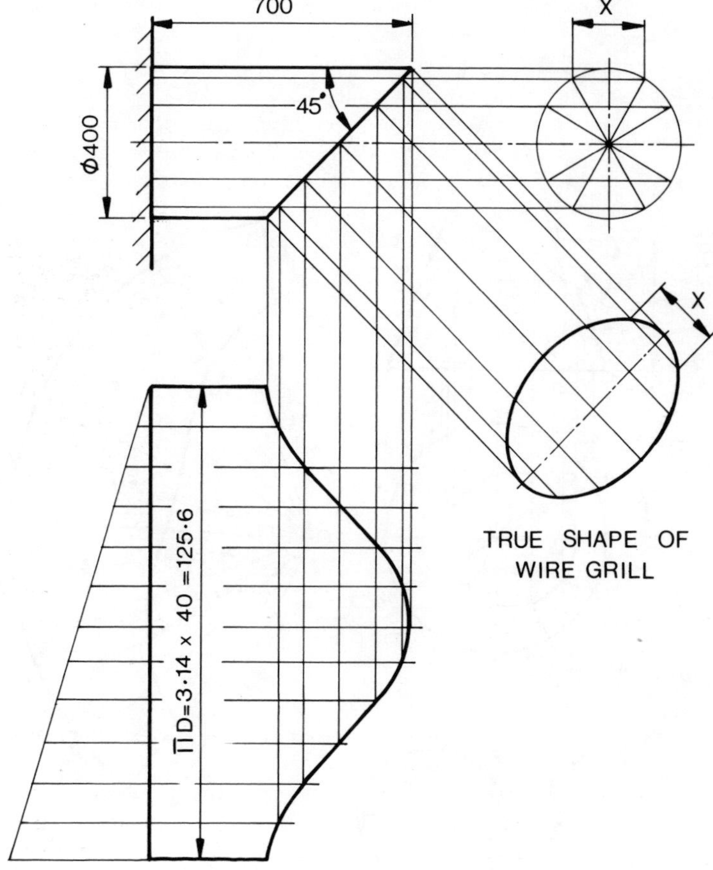

Fig. 6.12

and around the curve mark twelve chordal distances taken from the circumference on the plan view. Number the points and draw radial lines to the centre of the arc. It should be noted that the length of the arc on the pattern should be equal to the length of the circumference of the cone in the plan view, so this construction is only approximately correct, but acceptable within limits of normal error, in drawing.

Now note that all the radial lines to the apex in the front elevation are

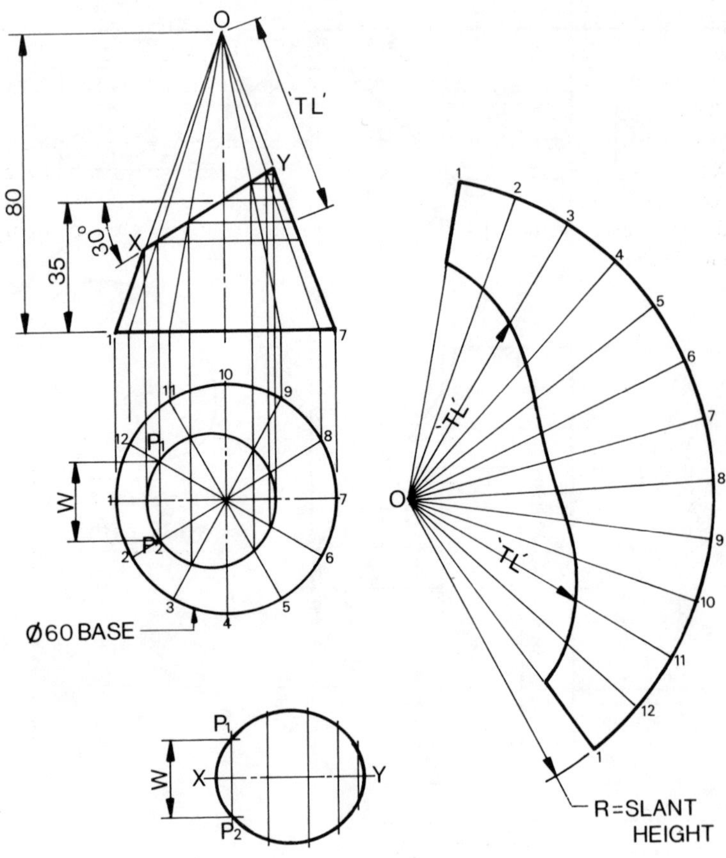

TRUE SHAPE OF ELLIPSE

Fig. 6.13 Note that the radius R in the pattern which is equal in length to the slant height of the cone is taken from the distance O1 to O7 in the front elevation.

equal in length but only lines O1 and O7 appear so in that view. Draw horizontal lines from the intersections of the sloping face with the radial lines across to line O7 in the front elevation. A typical true length shown as 'TL' for lines O3 and O11 is dimensioned on the pattern. Mark off the true lengths for all other points on the pattern and draw a curve through the intersections.

To draw the true shape of the ellipse, set out the major axis shown as XY. Two points on the ellipse are given as P_1 and P_2 and the width of the chord between them can be obtained from the true plan view. The distances between the vertical construction lines are taken from the sloping face in the front elevation.

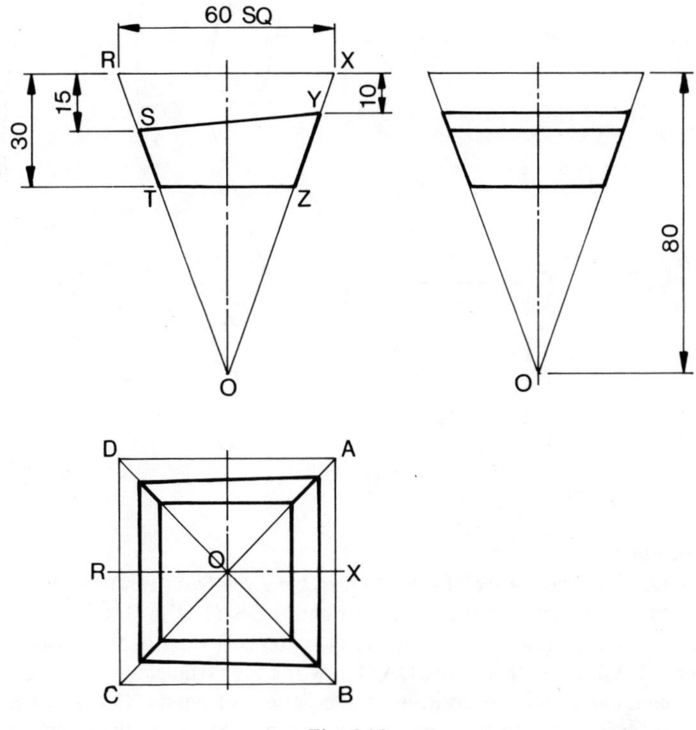

Fig. 6.14

In Fig. 6.15, the development for the sides and bottom of a wheelbarrow are shown drawn to scale. Copy the solution as a practical example of development work in everyday life.

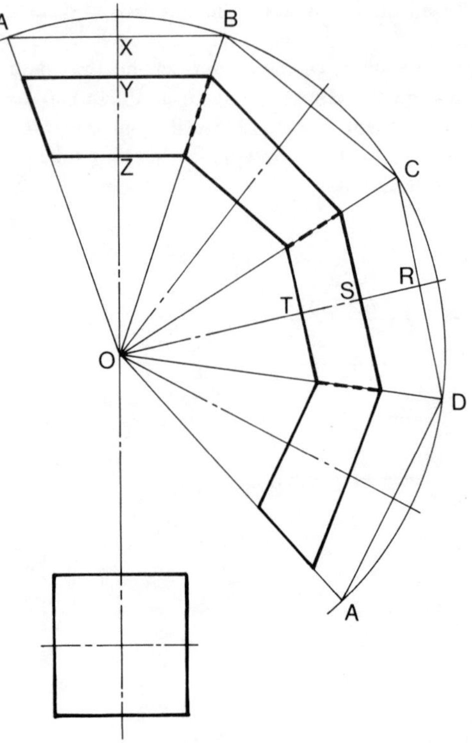

Fig. 6.15

Solution notes

Line OX is the centreline of side OAB in the given front elevation. Set out OX on the pattern and draw AB at right angles to OX. Draw an arc of radius OA and step off round the circumference a total of four distances equal to AB, letter them ABCDA. Draw the four triangles, including the centrelines and note the position on the pattern of points XYZ and RST. The distances between these points can now be taken from the given front elevation, measured along the sloping lines, and transferred to the pattern. Draw lines parallel to AB, also CD, through points Y, Z, S and T. Complete the pattern by adding the shapes in triangles OBC and OAD. The pattern shape for the bottom of the barrow can be copied directly from the true plan view since all lines in the plan are true lengths.

The construction for the pattern is a further example of conical development. To manufacture the barrow, the four sides and bottom would be cut on a single sheet of metal and arranged as indicated in Fig. 6.16. Redraw your pattern in this form and cut out the shape to check your construction.

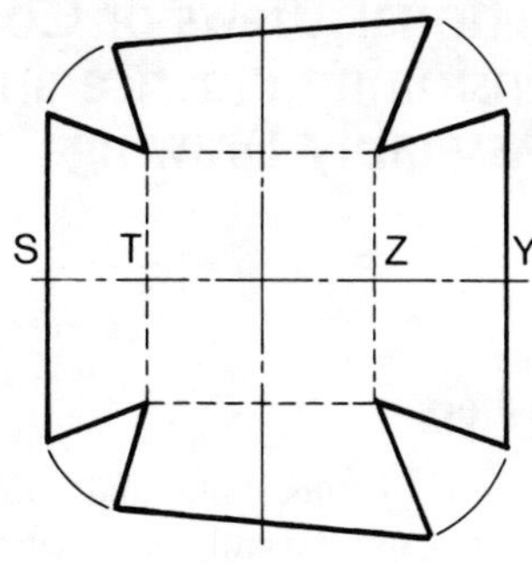

Fig. 6.16

7

Fastenings, Sectional Views of Components, Dimensioning Practice and Assembly Drawing

Nuts, bolts and washers

The approximate dimensions of nuts, bolts and washers are shown in Fig. 7.1. These items are used so frequently on assembly drawings that it is necessary to remember some approximate sizes so that the outlines can be produced rapidly. The dimensions on each of the components are given as proportions of the shank diameter of the bolt, dimension 'D'.

Bolts

Note that the length of the bolt is taken from the underside of the head and that the threaded portion is usually less than the length of the bolt. Instead of drawing the actual threads, a convention is used, and parallel lines represent the threads. The minor diameter of the thread is approximately ten to fifteen per cent less than the major diameter. The runout of the thread which is the point where the thread ceases is indicated by small lines at 30° to the bolt axis. The end of the bolt is generally finished with a radius or it may be chamferred at about 45° to remove the sharp corners and ease assembly. Both of these alternatives are illustrated in Fig. 7.1.

Nuts

An ordinary nut has the same hexagon as the bolt onto which it fits but its thickness is slightly larger. The nut is usually chamferred on one side only. A locknut is similar to an ordinary nut in plan view but its thickness

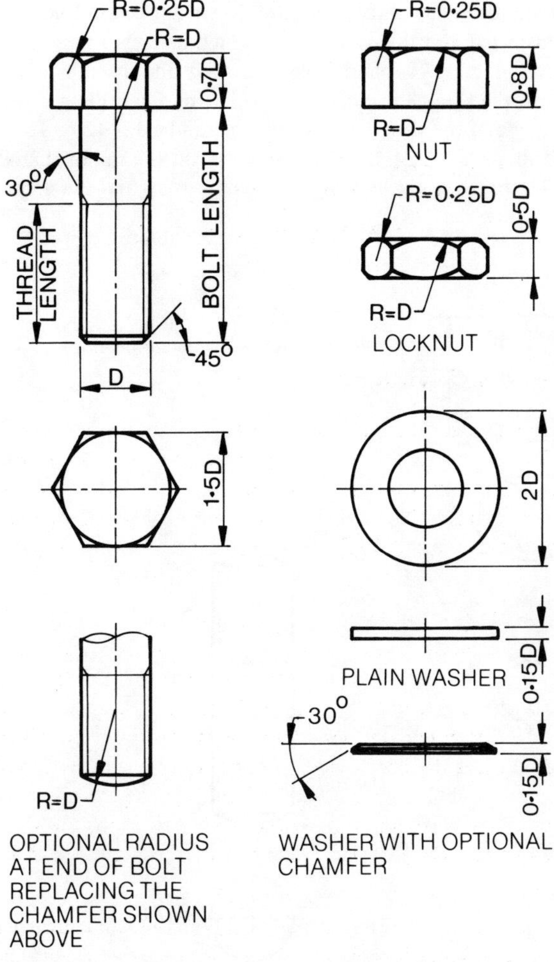

Fig. 7.1 Proportions of nuts, washers and bolts.

is approximately half the shank diameter of the mating bolt. Locknuts are also generally chamferred on both sides.

Washers

The outside diameter of a washer is approximately twice the shank diameter of the bolt and may have a plain or chamferred finish.

The standard thread conventions are shown applied in Fig. 7.2. A tapped hole is illustrated and this is a hole in which a screw thread has been cut. During manufacture, a blind hole is drilled and its maximum depth is measured to the tip of the cone which is left by the drilling operation. The included angle of the cone in the drawing standard is 120°. The size of the tapping drill is shown on the drawing equal to the minor diameter of the thread. The thread is cut by a tool known as a tap and the major diameter

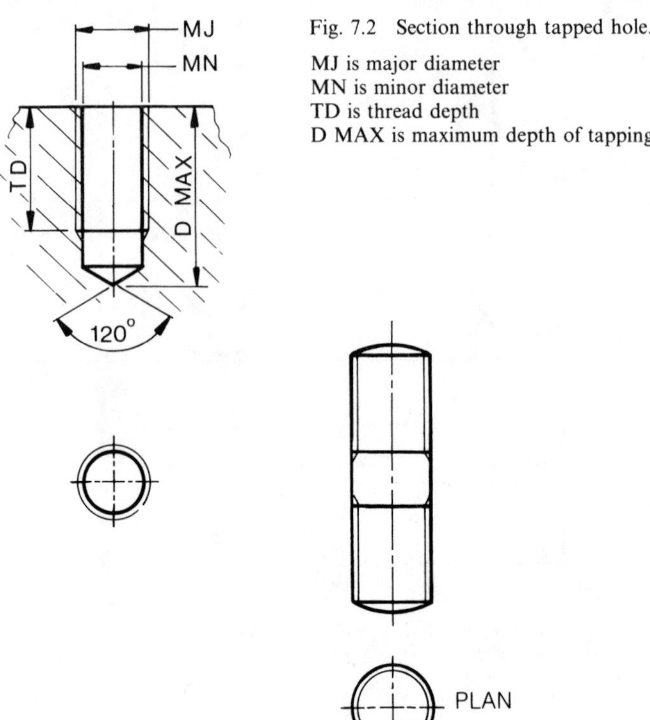

Fig. 7.2 Section through tapped hole.

MJ is major diameter
MN is minor diameter
TD is thread depth
D MAX is maximum depth of tapping drill

Fig. 7.3 Stud with plan view illustrating convention for a male thread.

of the thread is shown by additional parallel lines to the thread axis. The thread depth does not include the runout at the bottom which is indicated by short lines at 30° to the axis. In a sectional view which indicates a thread without an assembled bolt or stud it is conventional practice to cross hatch right up to the minor diameter. The plan view of a tapped hole shows two concentric circles indicating the major and minor diameters and with the female hole it is the custom to leave a small break in the major diameter circle.

Fig. 7.3 shows a stud which is a length of circular rod screwed at both ends. The screw thread may be cut with a tool known as a die. The plan view of the stud also indicates the standard convention for a male thread where the minor diameter circle is given a small break in order to distinguish it from the female convention.

A stud assembly is shown in sectional elevation Fig. 7.4. In sectional views it is normal practice not to cross hatch nuts, studs or washers so these components are seen in outside view. Notice that the cross hatching is only taken to the major diameter of the thread where the stud appears in the tapped hole.

When assembled, the stud is screwed into component 'B' so that the thread is level with the top surface of the tapped hole. The clearance hole in part 'A' permits the part to pass freely over the stud before the washer and nut are added.

Exercise Draw three views in orthographic projection of a nut to fit a 30 mm diameter bolt.

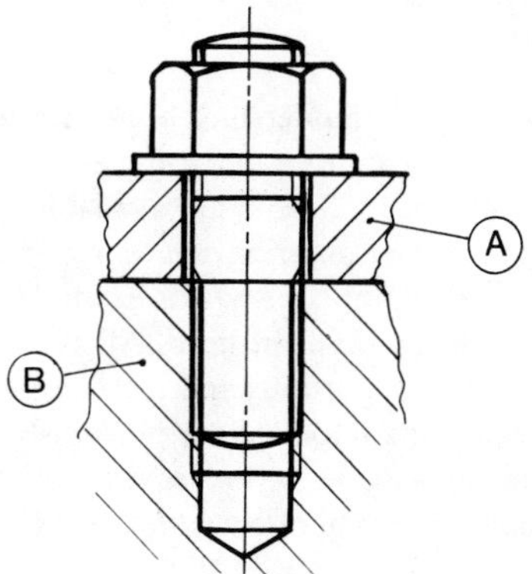

Fig. 7.4 Stud assembly.

Stage 1 Using the proportions illustrated in Fig. 7.1, commence the exercise by drawing a circle of Ø45 mm in the plan view. Add the six sides of the hexagon around this circle. Draw the major diameter of the thread shown above as Ø30 and in accordance with the thread conventions, a small break appears in the outline. Show the minor diameter of the thread as a complete circle of Ø27, which is calculated from the major diameter, less ten per cent. The height of the nut in the front elevation is 0·8D which will equal 24 mm. Project the corners of the hexagon up to the front elevation. From a point on the centreline, draw in the curve of radius 30, which results from the chamferring of the corners of the nut. Project vertically from the plan view the width of the chamfer circle as shown. Commence the end elevation ensuring that the width of this view is the depth of the plan view, namely 45 mm. Project across to the end elevation the depth of the chamfer.

Stage 2 Note that the end elevation of the nut has sharp corners since the chamfer circle touches the vertical centreline in the plan view at a tangent to the hexagon. The sharp corner of the nut along the centreline in the end elevation also terminates at the bottom of the chamfer curve. Complete the end elevation by adding the chamfer curves. The exact arc can be found by bisecting the chord as shown but the experienced draughtsman usually estimates the centre of arc and its magnitude. A little practice will be found to be of benefit at this point. Complete the front elevation by adding the chamfer curves on the receding faces of the hexagon and the straight lines at the edges of the chamfer.

Screws

The proportions of some widely used screws are given as these are also used on assembly drawings.

Socket screws

Socket screws with cap and countersunk heads are illustrated in Fig. 7.7. The screws are tightened in an assembly using a small hexagon shaped key which fits into the screw head. Screws are available in a variety of lengths and lengths of thread. When ordering screws it is necessary to quote the nominal size and this is the thread size, for example 10 mm, then add the screw and thread lengths and type of head. All 10 mm screws of the same type have the same head proportions and these are given below.

The cap type of screw fits into a recess previously counterbored on the surface of a component and the screw length does not include the head thickness. Countersunk screws fit into a countersunk recess of similar shape so that the screw is neat and flush with the surface. The lengths of countersunk screws include the head thickness.

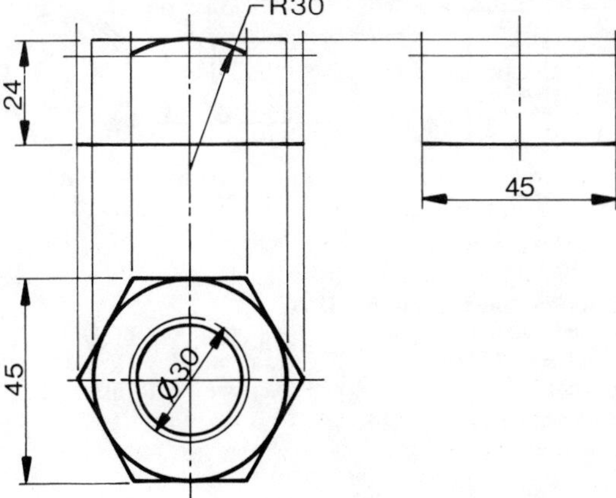

Fig. 7.5 Stage 1.

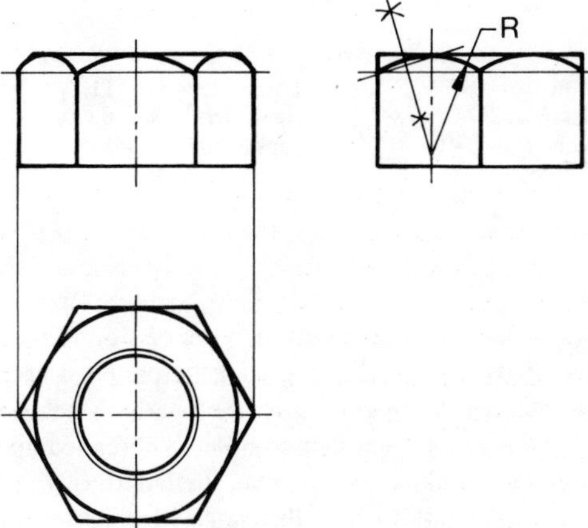

Fig. 7.6 Stage 2.

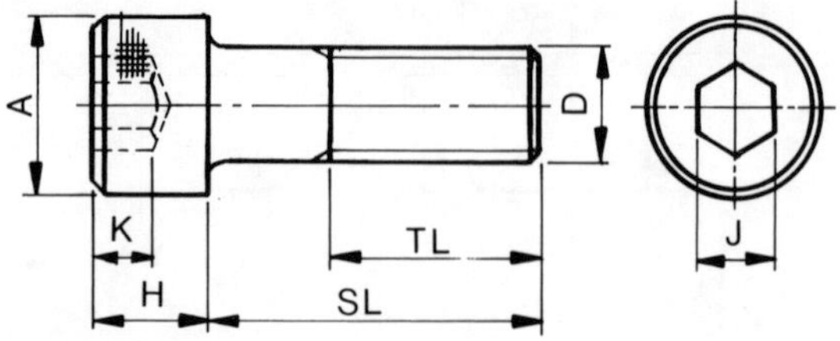

ISO metric socket cap screws to BS 4168.
Nominal thread size = D
Head diameter A = 1·5D
Key engagement K = 0·5D
Screw length = SL
Thread length = TL
Head depth H = D
Socket size J = 0·75D

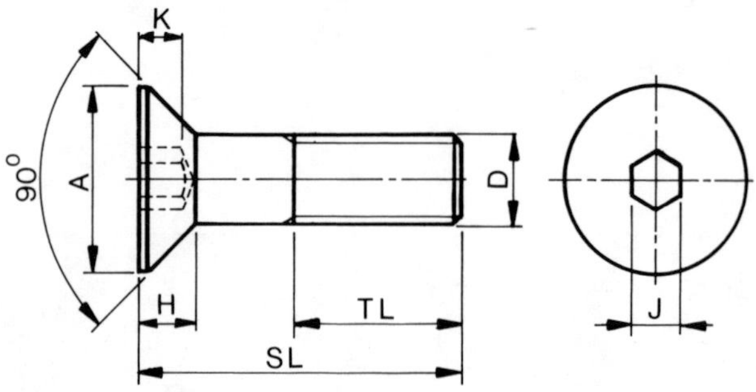

ISO metric hexagon socket countersunk head screws to BS 4168.
Nominal thread size = D
Head diameter A = 2D
Key engagement K = 0·4D
Screw length = SL
Thread length = TL
Head depth H = 0·6D
Socket size J = 0·6D

Fig. 7.7

Slotted screws

Approximate dimensions for the common types of slotted machine screws are given in Fig. 7.8. Note that the length of the panhead and cheesehead screws does not include the thickness of the heads. Machine screws are available in steel and brass, partially or completely threaded up to the head. The length of a countersunk screw includes the length of the conical part of the head. The countersunk screws illustrated here show a type used for flush finishes and a raised countersunk type often referred to as an instru-

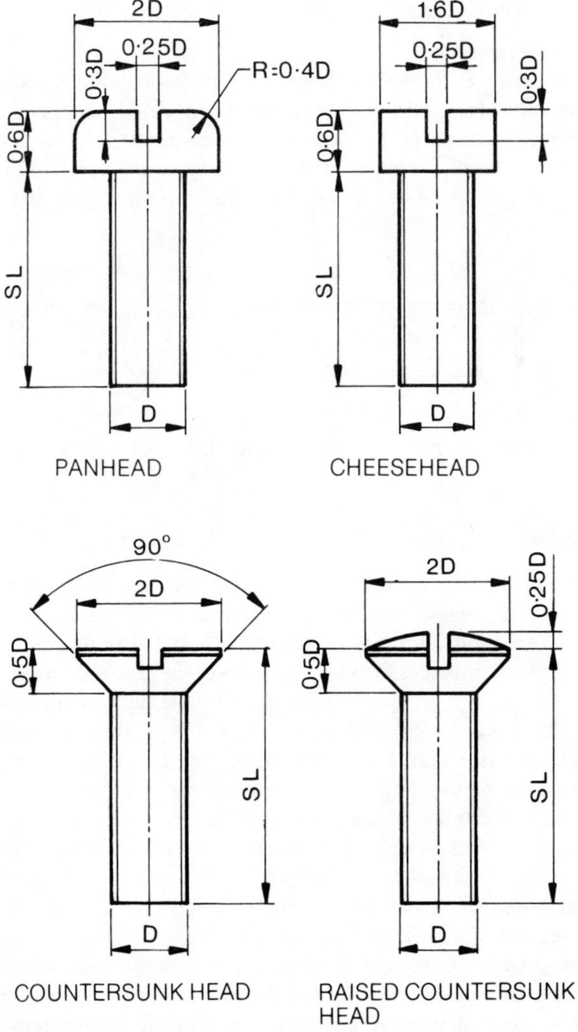

Fig. 7.8 Proportions of slotted machine screws. *Note:* All screw lengths shown as SL.

ment head screw. This type is often given a chrome-plated finish when used in prominent positions.

Posidriv screws

As an alternative to the slotted head screws, Posidriv screws which require a special type of screwdriver, are available with pan, countersunk and raised countersunk heads. Posidriv screws are often referred to as recessed head screws and a Posidriv countersunk head screw is illustrated in Fig. 7.9. The dimensions given in Fig. 7.8 can also be used to draw Posidriv screws.

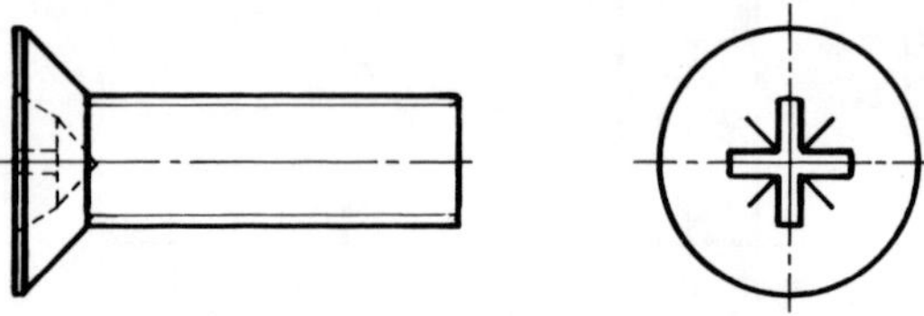

Fig. 7.9 Posidriv machine screw with countersunk head.

Sections

Hidden details of components can be indicated by dotted lines, but this practice can be somewhat confusing in complicated parts. As an alternative, we produce sectional drawings which show the detail along particular planes known as section planes. Take, for example, the block shown in Fig. 7.10. The block has a plain hole, a countersunk hole and a counterbored hole and if this is cut longitudinally and viewed in a direction square to the cut face, then we shall see a typical section. This component would be illustrated by the draughtsman as shown in Fig. 7.11.

In order to emphasise the actual surface along the section plane, cross sectioning, or cross hatching lines are added. The line thickness is 0·3 mm, the same as centrelines. Cross hatching lines must be spaced at equal pitches. Note that the rear corners of the countersunk and counterbored holes, not on the section plane, are outlined by solid lines.

In the plan view, the line of the section is clearly defined by the centre-line which has a thickened portion at each end. The arrows point to the half of the component presented in the front elevation. It is common practice to indicate under the sectional view where the section was cut, hence the statement 'Section AA'. The arrows in the plan are also lettered 'A' and 'A'. Obviously, in a more complicated component, more than one section

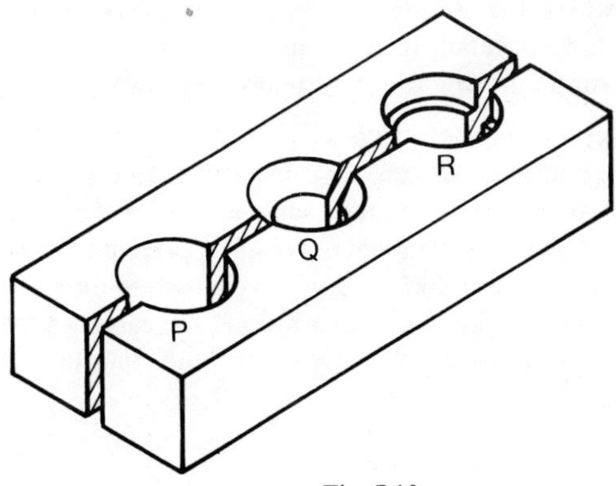

Fig. 7.10
P is a plain hole
Q is a countersunk hole
R is a counterbored hole

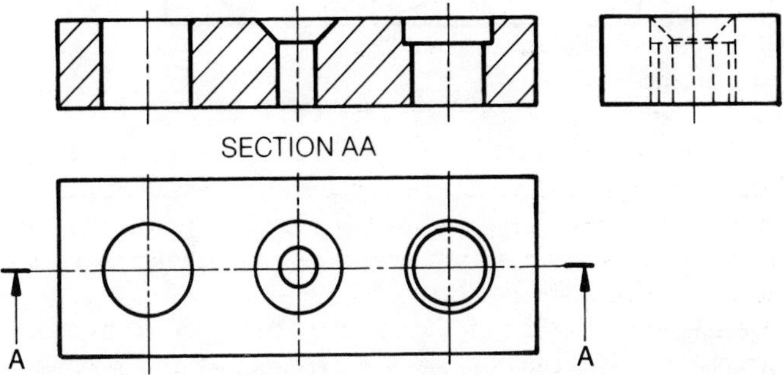

Fig. 7.11 *Note:* equally spaced cross hatching lines at 45° and statement Section AA beneath the front view also the indication of the section in the plan view.

may be drawn and each section is given different letters, since no doubt must exist as to where the section plane was taken. Note that the end elevation with its hidden detail, in this particular case, does not contain any useful information and would not normally be drawn.

Staggered sections

For convenience, we often draw staggered sections where features do not lie along the same straight line. If, for example, one of the holes in the block was displaced, then the component could be presented as shown in Fig. 7.12. The section plane is thickened where changes in direction occur. The elevations of course appear to be identical in each case, but the reader must take note of the information conveyed in both plan and elevation to appreciate the true form of the component.

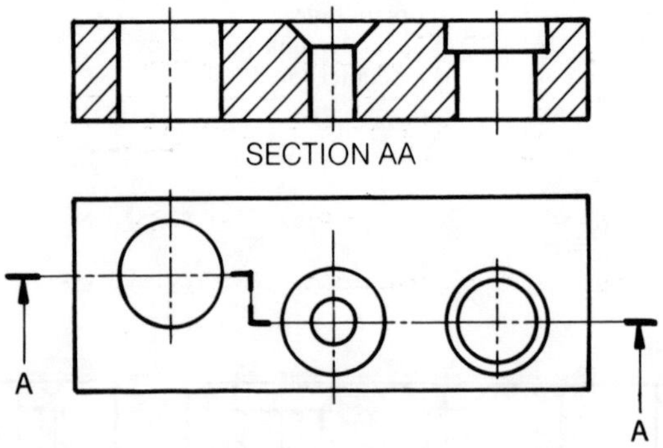

Fig. 7.12 An example of a staggered section.

Part sections

It is not always necessary to show complete sections and alternative methods of showing part sections are shown below. A part section is used to clarify minor details and in Fig. 7.13, an example is given to show a keyway in a shaft. In this case, it is not required to add a section plane and an irregular line is adequate to define the boundary of the section.

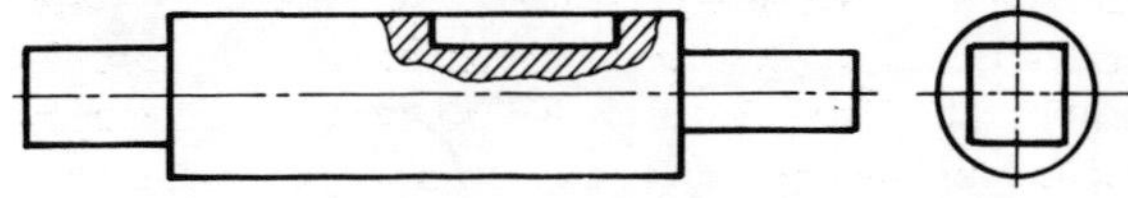

Fig. 7.13

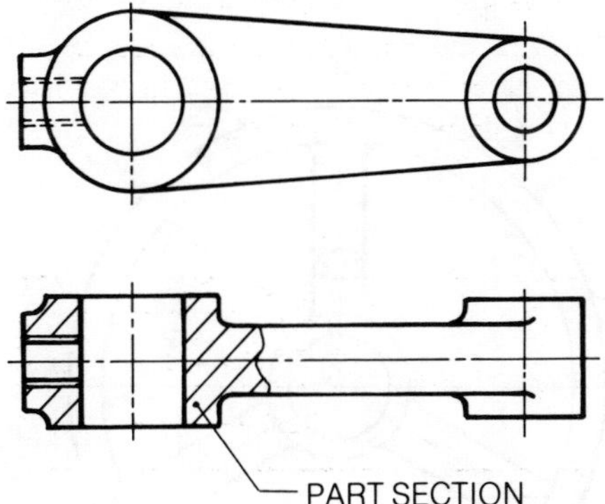

Fig. 7.14 Part section applied to the end of a lever.

Revolved sections

The shape of the web on the lever previously used in Fig. 7.14 cannot be determined from the limited information given on this drawing. The cross section can be indicated by a revolved section where a cut is made through the web and the resulting section turned through 90°. Fig. 7.15 shows a revolved section added to the drawing.

A further example of a revolved section is illustrated in Fig. 7.17 and this shows a convenient method of providing details of both bosses on the same end view. Note that the section is presented as though the boss 'B' and its centreline had been rotated about the centre of the plate onto the vertical centreline through the component. The section plane line is thickened where the change of direction takes place as well as at each end.

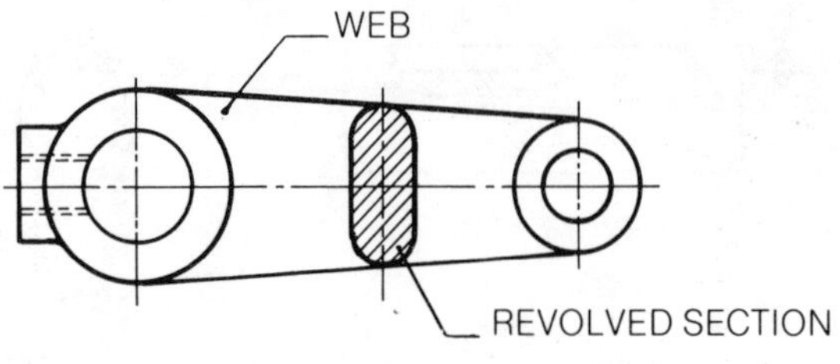

Fig. 7.15

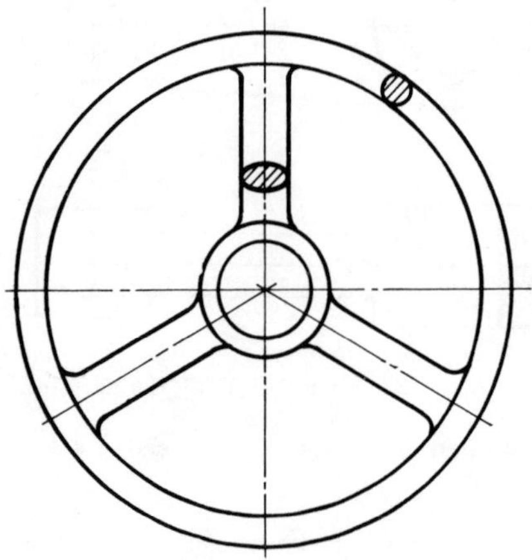

Fig. 7.16 Illustrating revolved sections taken through the spokes and rim of a steering wheel.

Removed sections

As the name implies, a removed section is positioned away from its normal location and this practice can be regarded as an alternative method of presenting some revolved sections. A removed section can be drawn at a larger scale, convenient for dimensioning purposes, and in such a case a section plane must be indicated together with the new scale used. The removed section should be drawn in projection with the cutting plane if possible, but on complicated detail drawings this is not always convenient and it is permitted to show this information elsewhere on the drawing sheet.

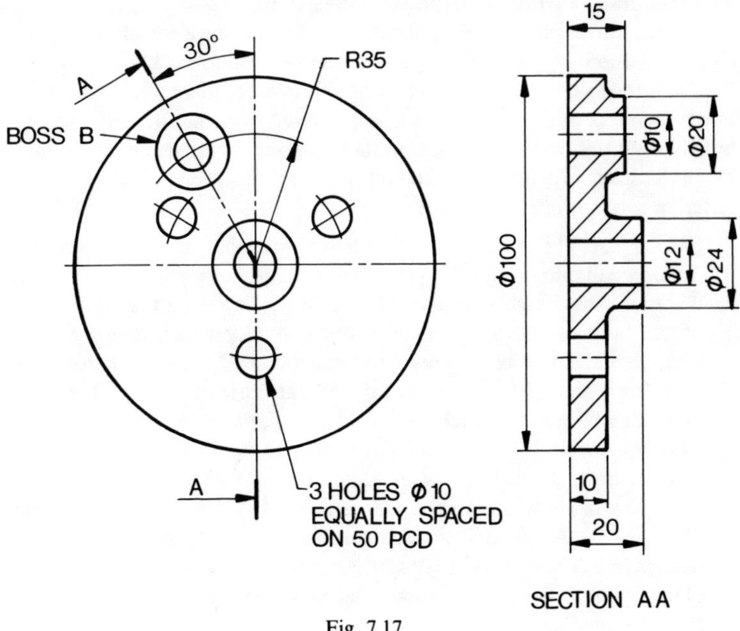

Fig. 7.17

A B

A B

SECTION AA SECTION BB

REMOVED SECTIONS THROUGH SPANNER HANDLE

Fig. 7.18 Illustrating removed sections AA and BB. The corners of the square section at AA have been cut away to give the octagonal shape at BB.

The treatment of thin webs in sectional views

There are three particular applications of sectioning which should be emphasised and these relate to thin structural supports, known as *webs*. Fig. 7.19 shows a component consisting of three cylinders, each with a hole bored through, and the cylinders are supported horizontally and vertically by relatively thin webs. Assume that the component has been manufactured by the casting process and is therefore one solid piece of metal. Three different sections are illustrated.

1 *Section 'AA'* This is a sectional view taken along the horizontal centreline. Theoretically, to obtain this plan view we must cut through metal which is shown cross hatched in plan 'D', but this view gives a false impression of the bulk of the object and suggests solidity which does not exist because of the webbed nature of the construction. Note that in the section 'AA', only the part of the vertical web cut horizontally by the section plane has been cross hatched and the two parts of the horizontal web, shown as areas 'X', have been left blank. Theoretically, the inner lines of these areas should consist of dotted lines but it is standard practice never to cross hatch up to boundaries defined by dotted lines, and hence a full line is always used for this particular application.
2 *Section 'BB'* The alternative situation is illustrated here, where the section plane completely passes through the thin vertical web and this is not cross hatched.
3 *Section 'CC'* The vertical section plane cuts across both of the webs and the complete area is cross hatched. The background information completes the view but the dotted lines which represent the drilled hole are omitted. It is standard practice not to show hidden details on sectional view, unless such information is not given on other elevations and plans.

   Redraw this component and the three sectional views to the dimensions provided.

Half sections

Symmetrical parts may be drawn half in full outside view and half in section as shown in Fig. 7.20.

Adjacent parts

Circumstances often arise where it is desirable to show an adjacent part close to a detail which has been drawn as a section and Fig. 7.21 gives a typical application of a cover fitting into a recess on a cylindrical component. The adjacent part should not be allowed to hide the section which must be regarded as the principal object of the drawing.

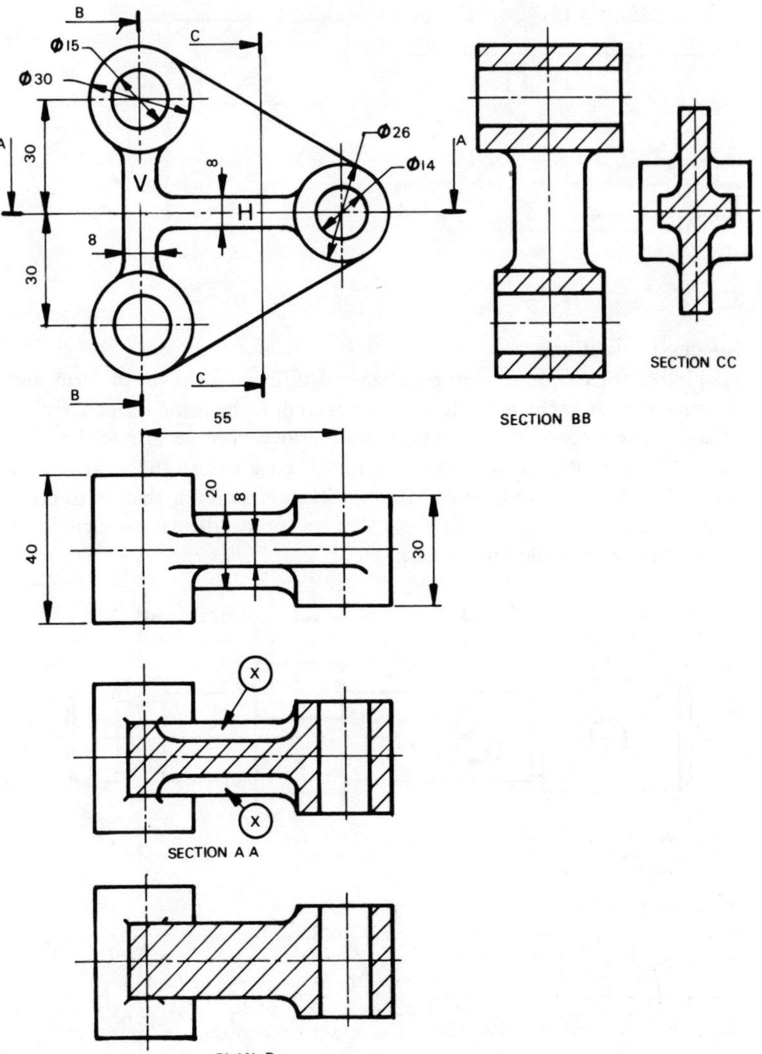

Fig. 7.19 V denotes vertical thin web and H denotes horizontal thin web.

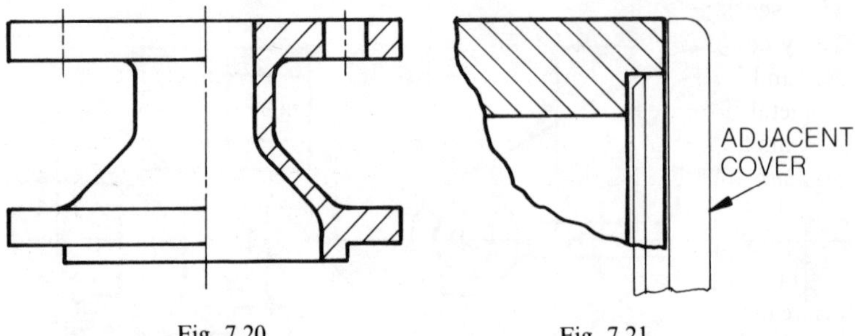

Fig. 7.20

Fig. 7.21

Successive sections

A spindle, for example, may have several different changes in form and the designer can indicate sections at various points by using the convention of successive sections. If the sections were drawn in projection to the right hand side, then they would take up a lot of space and so the arrangement given in Fig. 7.22 may be used. This illustration shows a shaft with cross holes, spanner flats, slots and holes that are perpendicular to each other which have been drilled in a recess.

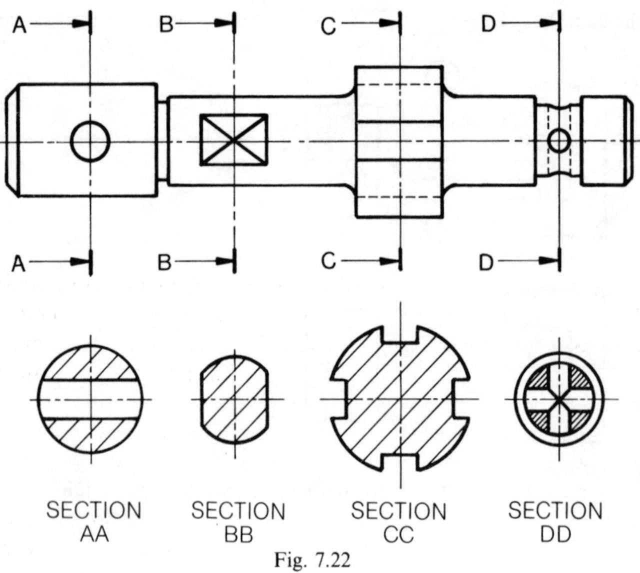

Fig. 7.22

Thin sections

Many components are manufactured from materials which are relatively thin and it is not possible to cross hatch sections. Typical examples would be metal boxes, covers, lids, brackets and thin gaskets. Where such components appear in sectional views it is normal to draw a thick line, but in the case of adjacent parts a small gap is left between them for clarity. Fig. 7.23 shows a cross section through a compound stanchion used in structural steelwork fabrications. This assembly consists of two metal plates fixed to an 'H' section joist. The gap left between the separate parts enables the reader to appreciate the outline of the three components forming the assembly.

Fig. 7.23

Components not drawn in section

It is the custom not to section many recognisable components even though they may be positioned on an assembly along the cutting plane. Nuts, bolts, washers, studs, rivets, pins, keys, spindles, balls, rollers, cotters are typical parts which remain unsectioned because of their clear external profiles and in the illustrations relating to machine drawing which follow, there are many examples.

Dimensioning

Technical drawings are used as plans to build or manufacture and they should contain a complete set of working instructions for the craftsman. Dimensions define features and outlines which are formed by simple geometrical characteristics, i.e. diameters, lengths, tapers, angles etc., and these items need to be located and positioned with each other.

Many standards exist relating to dimensioning and tolerancing and a book could be written on the subject alone, so in the space available here we can only introduce the reader to the basic principles.

Normally, there should be no more dimensions than are necessary to describe the end product. The dimensions are provided for the reader of the drawing and it should not be necessary to deduce a dimension from other dimensions nor scale the drawing for information. Any particular feature or location should be dimensioned once only.

It is sometimes desirable to add auxiliary dimensions which give useful information, for example, an overall approximate length. Auxiliary dimensions do not control manufacture or acceptability in any way but are sometimes convenient to know when marking out material before work commences. Auxiliary dimensions are shown on drawings between brackets.

A functional dimension, determined by the designer, is one which has a direct bearing on the operation of the assembly or the form of a single component. Sometimes these dimensions are subject to an extra tolerance. It is not possible to manufacture to exact dimensions and the designer normally gives a 'tolerance' which can be regarded as a deviation from an exact figure and this will ease production but still allow correct functioning of the component to take place. Tolerances on separate dimensions are not applied indiscriminately or too closely since the production process involved may prove difficult or uneconomic. If a given dimension was quoted as 25·00 ± 0·1, then the reader would deduce that an acceptable length would be between 25·1 mm and 24·9 mm. Any length between these two figures would be expected to function in the same way in an assembly. The length 25·1 is referred to as the high limit and 24·9 as the low limit. In this case, 25·00 is known as the nominal dimension and 0·1 the tolerance.

In many cases where a high degree of precision or accuracy is not required, then a general tolerance note can be quoted on the drawing. A typical note can be included in the drawing format or a boxed statement added as follows:

EXCEPT WHERE STATED, A GENERAL
TOLERANCE APPLIES OF ± 0.2

GENERAL MOULDING
ALLOWANCE ± 0.4.

The tolerance in the box will be assumed to be applicable to every dimension which is not toleranced with another figure.

Dimensioning terms

Fig. 7.24 shows a simple component where common terms used in dimen-

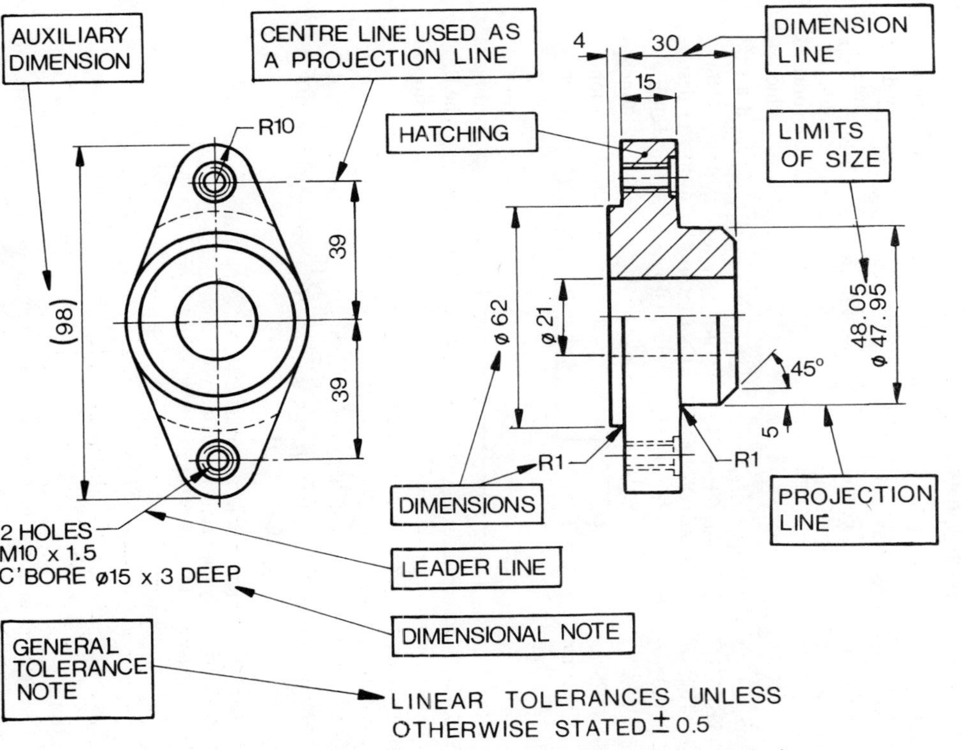

Fig. 7.24 Typical half section to illustrate common terms used in dimensioning.

sioning have been applied. The example shows a typical engineering detail drawing presented in half section. Half of the end view is shown in sectional form indicated by the hatching.

Leader lines

A leader line is used to indicate where notes and dimensions apply on a drawing. A leader is a continuous thin line and there are three particular applications illustrated in Fig. 7.25.

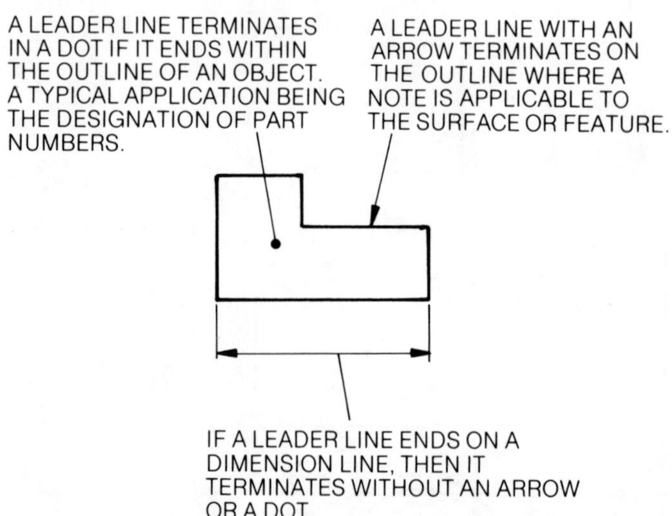

Fig. 7.25 Termination of leader lines. Note that the leader line should be drawn nearly normal to the surface to which the note applies and must not be parallel to other dimension or projection lines.

Datums

A dimension is used during the course of manufacture and the manner in which a component drawing is dimensioned often influences the sequence of marking out the job, production and final inspection. In the case of the simple component in Fig. 7.26, the edges XY and YZ would be finished square to each other and the position of the hole and the corners marked out from these two edges. XY is known as the vertical datum and YZ the horizontal datum. The dimensions indicated by the letter 'B' indicate the size of features and locate them from the datum edges. The dimensions 'A' are known as overall dimensions and as the name implies gives the overall length, width or height of a component. A datum need not be an edge and is often an axis. It is also not necessary to mark a datum edge or axis on a drawing as a general rule.

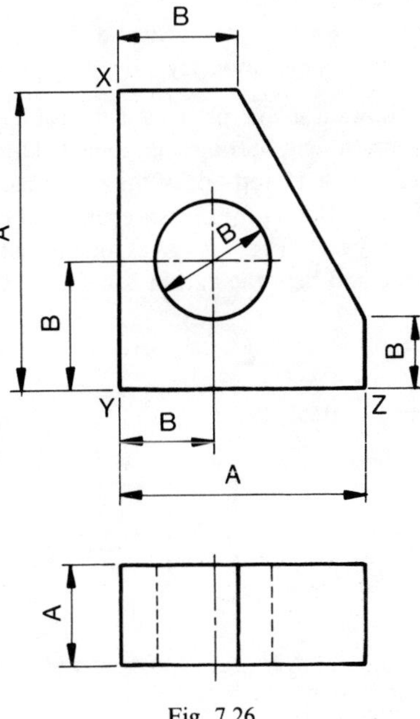

Fig. 7.26

Dimensioning rules

Dimensioning should be undertaken with care and the following guide lines memorised.

1. Position dimensions away from the view if this is possible.
2. Dimension a feature only once where its contours are clearly shown.
3. Maintain uniform spacing between dimension lines as this will give a neat tidy appearance to the drawing. A spacing of about 12 mm will give an adequate distance to add tolerances, if required.

4 Take particular care with arrowheads keeping them of uniform size.
5 Position the dimension figure centrally between the extension lines.
6 Dimension a feature on a view where it is shown with a full outline, in preference to dimensioning hidden detail with a dotted line.
7 Note that extension lines extend slightly beyond dimension lines and that there is also a small gap between the extension line and feature.
8 Extension and dimension lines are thin continuous lines.

Dimension lines must show a clear origin and termination but the method adopted varies with the type of industry.

Case 1 Fig. 7.27 shows a gauge plate as a typical example of a factory manufactured mechanical engineering component. Note that closed filled arrowheads are used. The left hand side of the plate, marked with the letters XY, has been used as the datum for measurement and each of the separate dimensions is equally spaced. Fig. 7.28 gives an alternative method of progressive dimensioning and here the datum is defined by the filled-in dot.

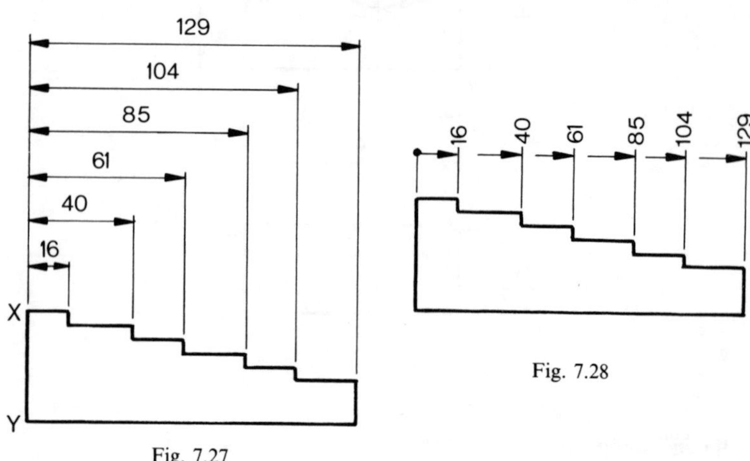

Fig. 7.27

Fig. 7.28

Case 2 In building and civil engineering drawings, a different technique is adopted. Open arrowheads are used and Fig. 7.29 shows a typical application of superimposed progressive dimensioning. In this example short lines form barbs at 90°. The datum or origin is indicated by an open circle. Note that the figures are near to the arrowheads.

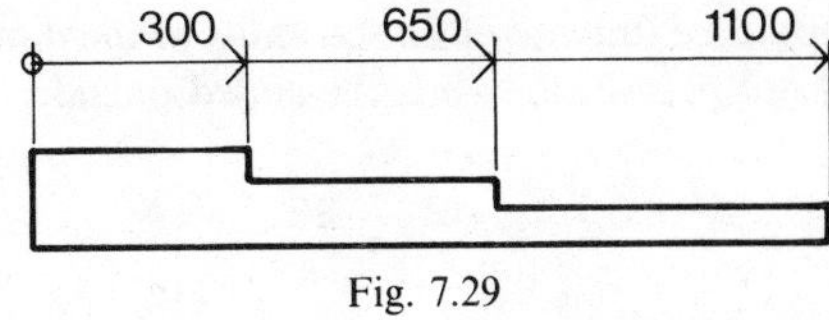

Fig. 7.29

Oblique strokes can also be used to show an origin or a termination and Fig. 7.30 illustrates this case. The term 'chain dimensioning' is often used in connection with the application of successive dimensions along the same line.

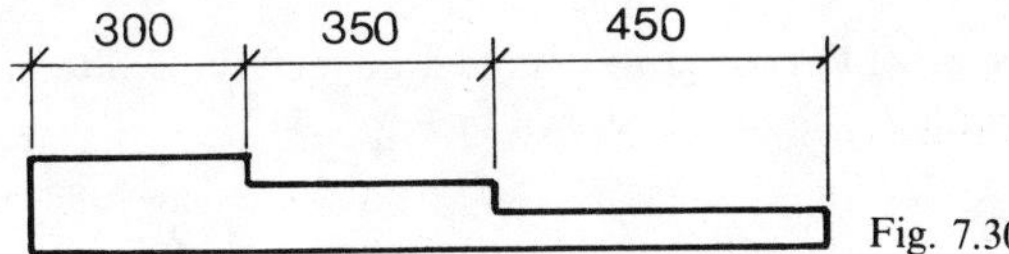

Fig. 7.30

If a feature is drawn with a common centreline, then the drawing implies that symmetry is required and that the workshop will manufacture on this basis. Closer tolerances can be applied and the geometric tolerance of symmetry is one of many topics in a British Standard on general geometric tolerances.

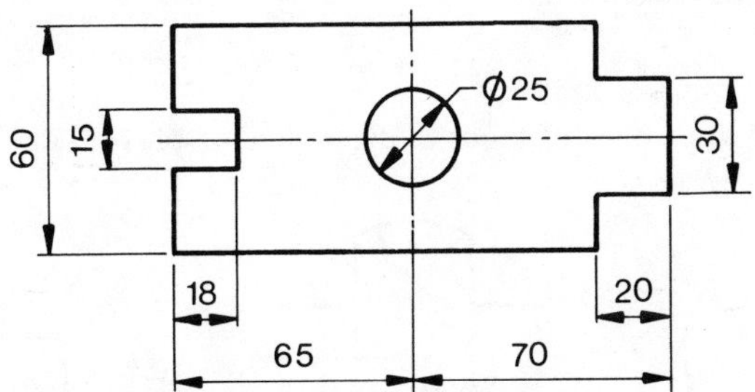

Fig. 7.31 Illustrating three features symmetrically spaced about a common horizontal centreline.

Dimension lines are always shown unbroken, even where the feature to which they refer may have been reduced in length.

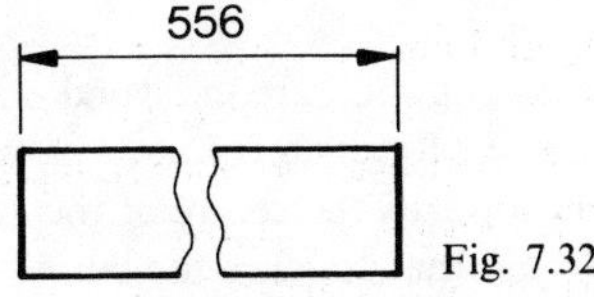

Fig. 7.32

Arrowheads should be drawn within the limits of the dimension line, but where space does not permit, they may be added outside.

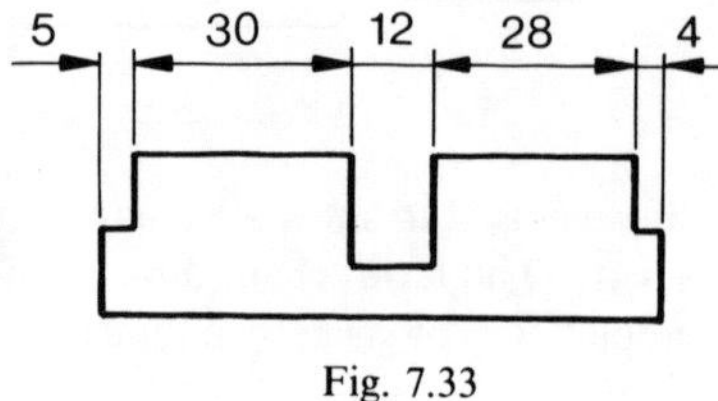

Fig. 7.33

The symbol Ø is used to denote a diameter and alternative methods of dimensioning a circle are shown in Fig. 7.34.

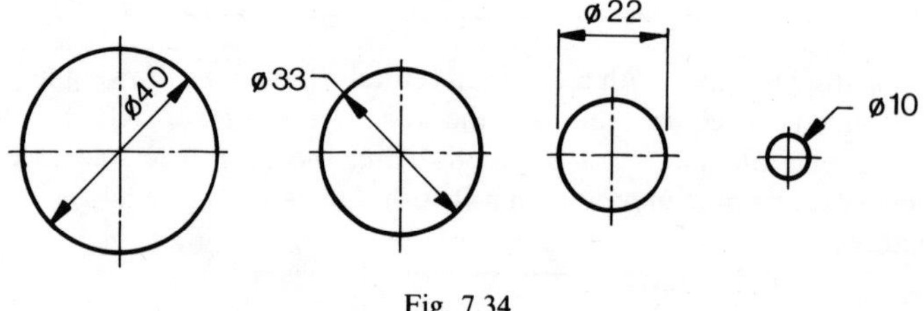

Fig. 7.34

Fig. 7.35 gives three methods of dimensioning spherical diameters.

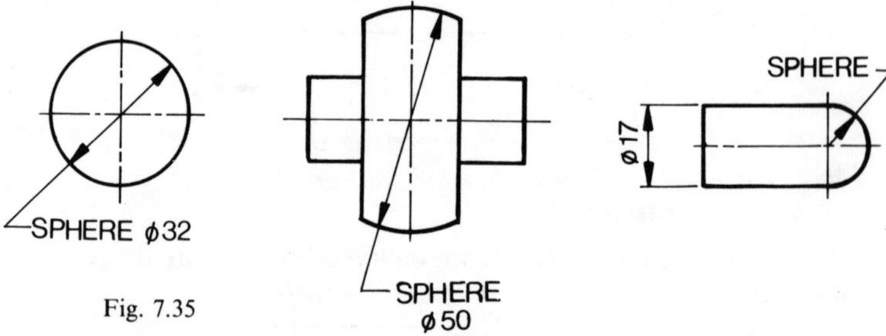

Fig. 7.35

Methods of dimensioning radii are indicated on Fig. 7.36. The abbreviation for radius is the letter R which is followed by the size. The dimension line is always directed towards the centre of the arc and should be finished with one arrowhead only that touches the curve.

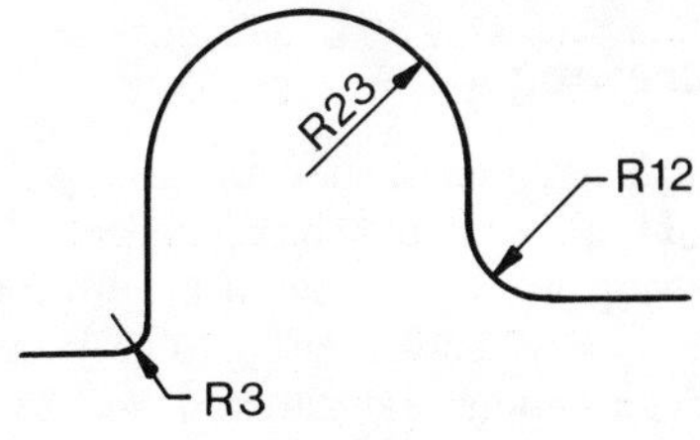

Fig. 7.36

Dimensioning screw threads

Fig. 7.37 shows a typical male thread where dimensions likely to be required include the following:

(*a*) The bolt length shown here is 63 mm.
(*b*) The thread length is 36 mm.
(*c*) An undercut at the end of the thread is 4 mm wide.
(*d*) The thread quoted is a 20 mm Metric thread with a pitch of 2·5 mm manufactured to a 6g BS tolerance.

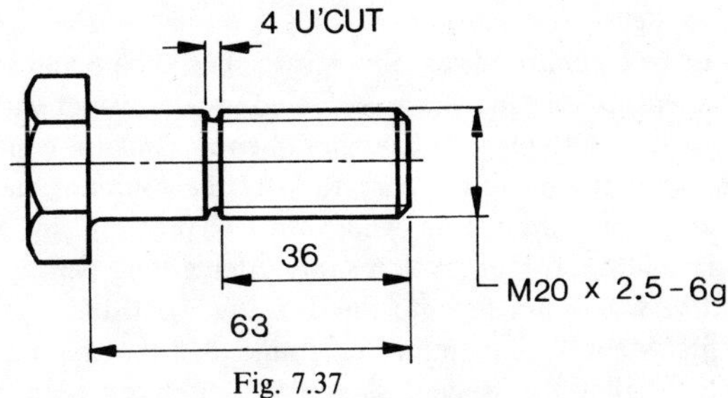

Fig. 7.37

Fig. 7.38 illustrates a typical tapped hole drawn as a section. Note that the maximum depth of hole before tapping is given to the tip of the cone left by the drill. The length of the full thread is quoted.

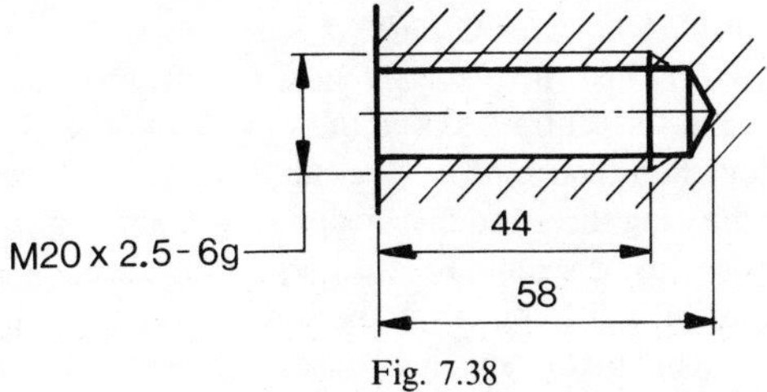

Fig. 7.38

The assembly drawing

This type of drawing is required to show the arrangement of parts so that the unit can be fitted together in the correct order.

The assembly drawing incorporates all of the principles which have previously been described in this book relating to projection, sections, dimensioning, draughting conventions and standards to BS308:1972 and for this reason most Examination Boards consider it necessary to test the student with a separate examination paper in this aspect of technical drawing. The following problems are reproduced by kind permission of various National Examination Boards and they have been selected to illustrate and emphasise important details. Solutions are required in either first or third Angle projection and the student is generally free to choose, but he must state the projection angle used.

In order to assist the reader, the first few examples are followed by our interpretation of the solution together with notes regarding the construction of the drawing. Please copy these solutions and try to appreciate the reason for the existence of every line.

It must be emphasised that much care should be taken with the drawing layout to ensure that the views presented are consistent with the projection angle selected. Both projection angles of course possess equal merit. Experience shows that it is all too easy to start the solution, and very tempting to do so in an examination where time is pressing, by working on one particular view and then to find sadly after a long period has passed that the next view will not fit onto the drawing sheet due to poor positioning of the first view. It is therefore recommended that before any drawing is undertaken, that the student sketch to a reduced scale the solution he proposes to offer, bearing in mind the projection. The sketch does not have to be complete in every detail, often the shape or silhouette being sufficient to enable suitable spacing of the views to be arranged. A reasonable sketch for this purpose to clarify the problem in the mind of the candidate can generally be done in a few minutes. Clearly, if two views are to be positioned across a sheet of paper then three spaces must be left, one on either side of the sheet and one in the centre between the views. Add the overall width of the two views together, then subtract the sum from the distance between the borders of the drawing sheet and finally divide by three to give the width of each space. Repeat this procedure for the vertical distances. Some adjustment may be made for a title block at the bottom right hand corner in some instances. Remember that spacing does not have to be exact to the last millimetre but marks are always awarded for good draughtsmanship

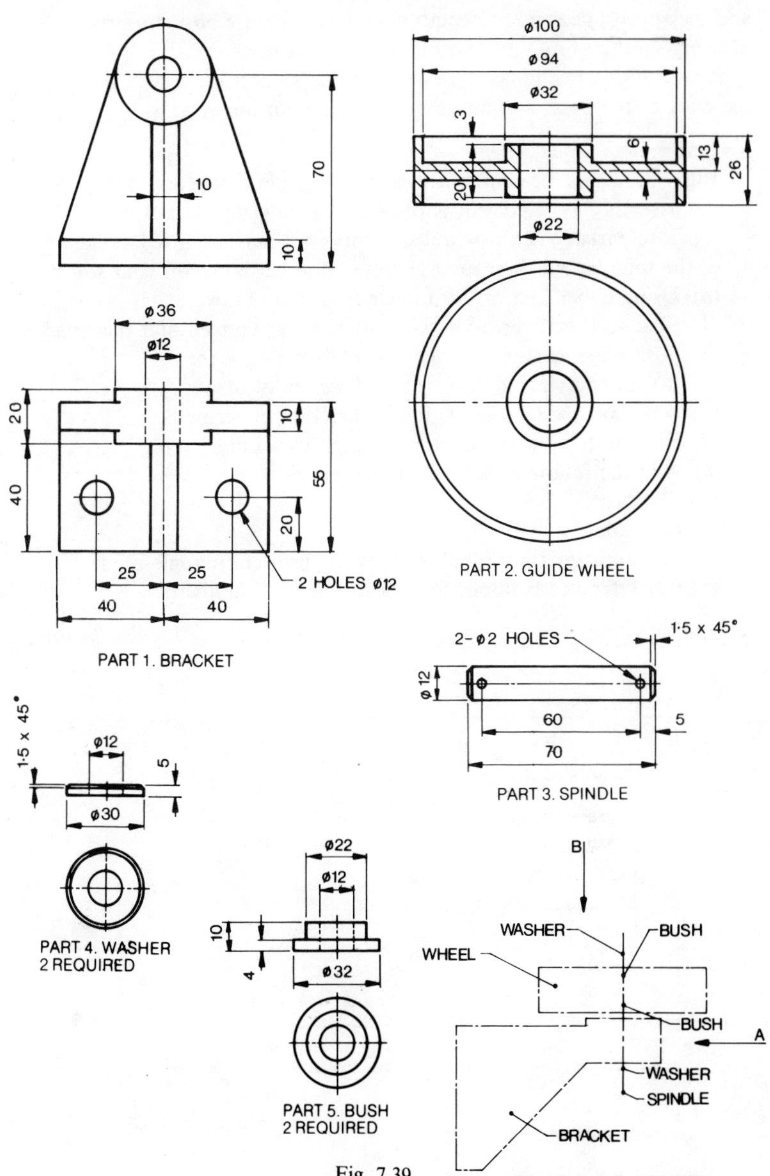
ø100
ø94
ø32
3
6
13
26
20
ø22
70
10
10
ø36
ø12
20
10
40
55
20
25
25
2 HOLES ø12
40
40
PART 1. BRACKET
PART 2. GUIDE WHEEL
2–ø2 HOLES
1·5 x 45°
ø 12
60
5
70
PART 3. SPINDLE
1·5 x 45°
ø12
5
ø30
PART 4. WASHER
2 REQUIRED
ø22
ø12
10
4
ø32
PART 5. BUSH
2 REQUIRED
B
WASHER
BUSH
WHEEL
BUSH
A
WASHER
SPINDLE
BRACKET
ASSEMBLY ARRANGEMENT

Fig. 7.39

and the appearance and presentation of a solution are both important. The drawing sheet can now be laid out with confidence.

The solutions to the last questions in this section have been printed at the back of the book so that you can check your answers.

1 *Sliding door guide roller*

Fig. 7.39 shows the component parts of a guide roller for a sliding door. An assembly arrangement is provided to indicate the positions of the separate parts. When assembled, 2 mm split pins are fitted at each end of the spindle, but these are not required to be shown on your solution.

(*a*) Using either first or third angle projection draw:

(i) a sectional elevation with the parts assembled and the bracket positioned as shown in the assembly arrangement.

(ii) an end view looking in the direction of arrow A.

(iii) a plan view looking in the direction of arrow B.

Hidden detail is required in the plan view only.

(*b*) Add the following to your drawing:

(i) title.

(ii) scale.

(iii) appropriate symbol to show the projection angle used.

(iv) reference balloons to indicate the part numbers.

City and Guilds of London Institute

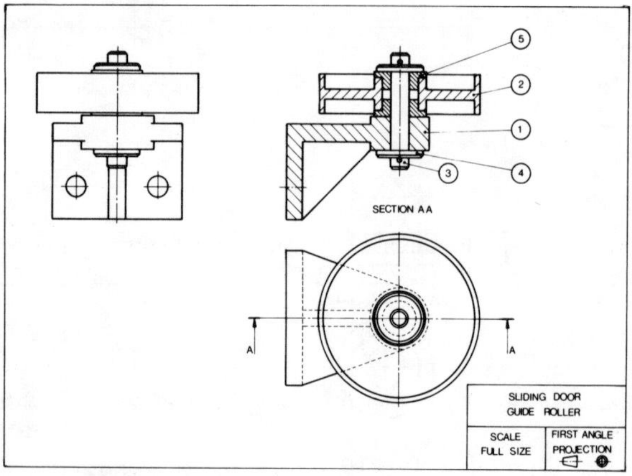

Fig. 7.40

Solution notes

Notice that the section plane in the plan view passes through a thin web on the bracket and the web is left without cross hatching in the sectional elevation. It is customary not to section washers and the spindle so these components are drawn in the sectional view as they normally appear. In order to indicate the position of each part in the assembly, balloons have been added with the part number in the centre and the leader lines joining the balloons to the drawing terminate with a dot on the part itself. Although there are two washers and bushes, it is only necessary to add balloons to one of each of the parts. The cross hatching on the bushes is drawn closer together as these are smaller parts. A sectional view should state where the section is taken from and hence the statement 'Section AA' and the section plane and arrows pointing in the right direction are included.

A solution in third angle projection would have had the three views arranged as shown below. The third angle symbol would also be added.

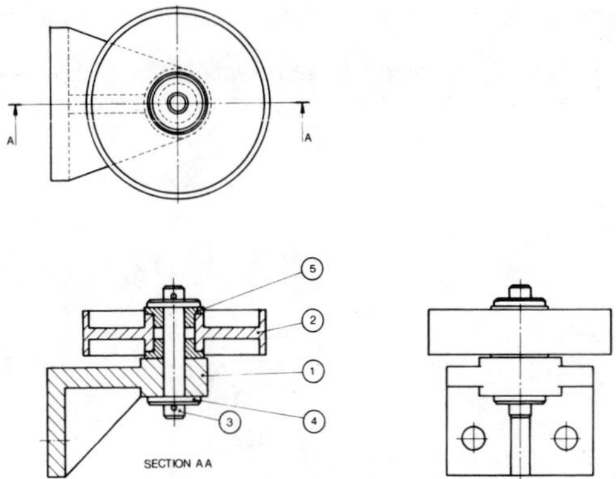

Fig. 7.41 Alternative arrangement of views in third angle projection.

2 *Electrical coil winding machine*

A pictorial view and details of each component part of a coil winding machine are shown. A clip which fastens the wire to the Coil Former has not been included.

Draw, full size, either in first or third angle projection the following views of the Unit completely assembled:

(i) A sectional front elevation taken on the cutting plane YY shown in body detail Part E.

(ii) A sectional end elevation taken on the cutting plane XX shown in body detail Part E.

(iii) A plan projected from view (i).

Hidden details are not required in any view.

Use your own judgement to determine the size of any dimensions not given.

A dimension shown as M10, for example, should be understood as M means metric thread. 10 means diameter of the shaft or hole in mm.

Print in the title – COIL WINDING MACHINE – size of letters to be 7 mm high.

Print in the scale and the system of projection used. Size of letters to be 5 mm high.

Put in the following dimensions:

(*a*) the overall height of the assembled machine.

(*b*) the length between the outside ends of the coil former locking nuts.

(*c*) the length between the inner faces of the end bearing brackets.

Associated Lancashire Schools Examining Board

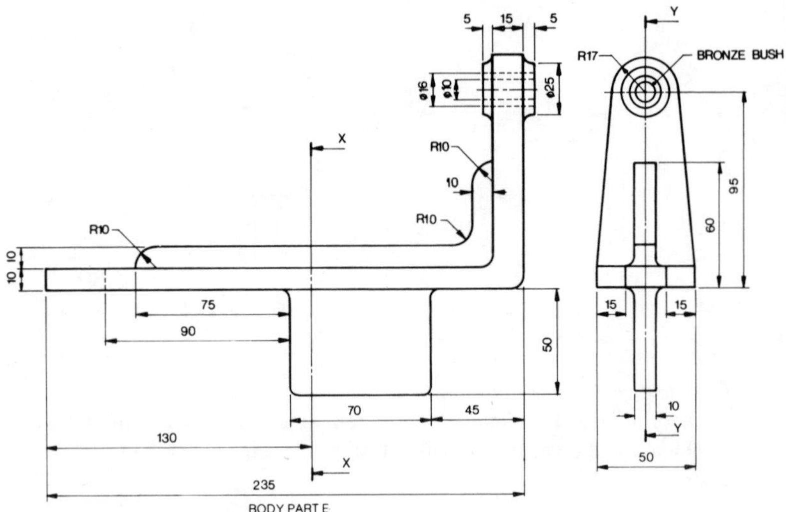

Fig. 7.42 Electrical coil winding machine. First angle projection.

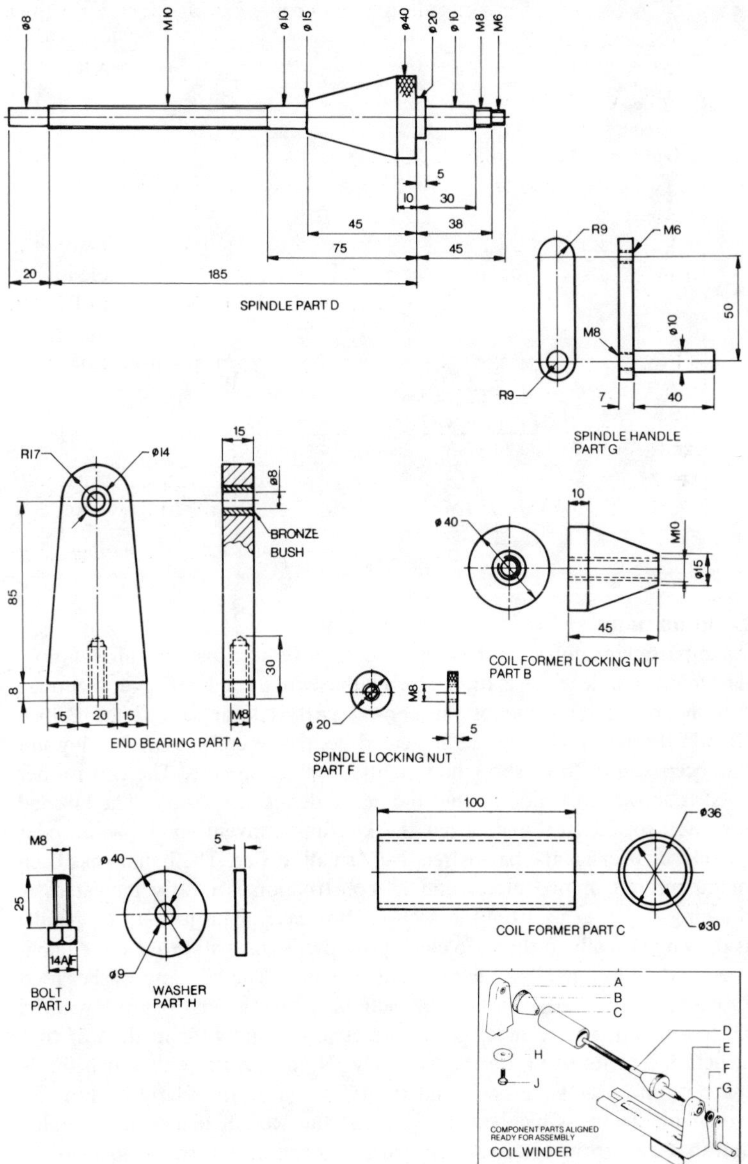
ø8
M10
ø10
ø15
ø40
ø20
ø10
M8
M6
5
10
30
45
38
75
45
20
185
SPINDLE PART D
R9
M6
50
ø10
M8
R9
7
40
SPINDLE HANDLE
PART G
15
R17
ø14
ø8
BRONZE
BUSH
85
30
8
15
20
15
M8
END BEARING PART A
10
ø40
M10
ø15
45
COIL FORMER LOCKING NUT
PART B
M8
ø20
5
SPINDLE LOCKING NUT
PART F
100
ø36
ø30
COIL FORMER PART C
M8
5
ø40
25
14AF
ø9
BOLT
PART J
WASHER
PART H
A
B
C
D
E
F
G
H
J
COMPONENT PARTS ALIGNED
READY FOR ASSEMBLY
COIL WINDER

Solution

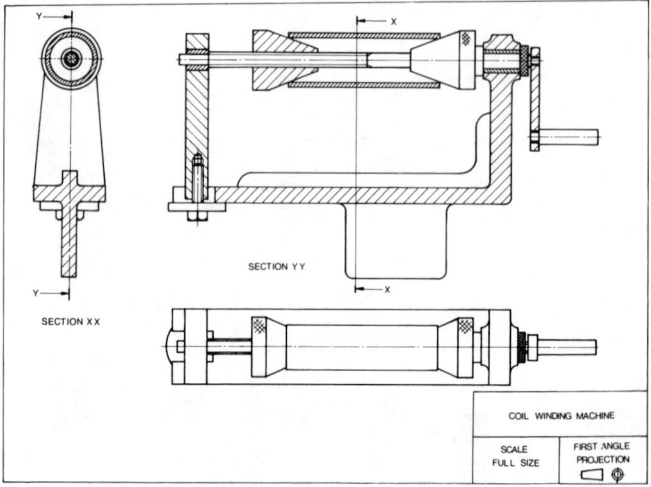

Fig. 7.43

Solution notes

Take particular notice of the components to be sectioned in this solution. The main spindle which is turned on a lathe from a solid bar is not sectioned but the parallel portion of the conical part is knurled and the British Standard convention must be indicated here. The coil former locking nut has been sectioned to show how it fits over the spindle. The coil former itself is shown in section so that the inside details are visible. The knurled nut used to lock the spindle has the knurling convention shown all over its surface because the part is relatively small in size. The handle has been manufactured in two pieces and its construction is clearly indicated by drawing the lever part only in section. For ease of projection the handle is drawn vertically in the six o'clock position as this will provide the maximum amount of detail in the sectional elevation. The bolt and washer used to clamp the end bearing are not sectioned, but the end bearing with its bush are sectioned, as again, this is necessary to show the method of construction and its fixing on the assembly. Note that the web which forms the bottom part of the body and the rib above it are relatively thin and would not be cross hatched. Ensure that the section planes are included and that the arrows are consistent with the projection used in the solution.

Alternative solution in third angle projection

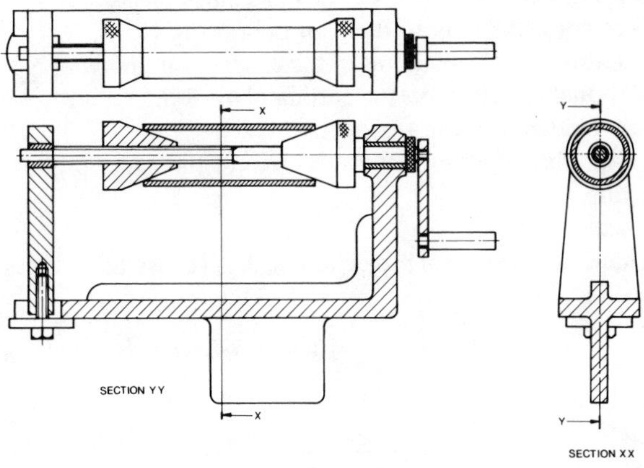

Fig. 7.44 Position of views in third angle projection.

The following problems are included for additional practice in drawing components and assemblies and in each case represent about one hour's work. All solutions are given at the back of the book and can be compared with your answers. Our solutions are given in first angle projection but the equivalent third angle solutions can easily be derived if the solutions for the sliding door guide roller and coil winding machine are studied and the rearrangement of views checked.

3 *Draw bolt*

The detail drawings show parts of a draw bolt of the type used to bolt heavy doors.

Drawing No. 1 shows the plan and two elevations of the body and catch plate mounted 15 mm apart.

The 18 mm diameter shaft shown in drawing No. 3 slides through the body and into the catch plate.

The M8 setscrew shown in drawing No. 2 screws into the shaft and acts as a handle and locking device.

With the parts assembled and in the locked position, draw full size the following views in either first or third angle projection:

(*a*) a plan corresponding to the plan in drawing No. 1.
(*b*) an elevation corresponding to the elevation in drawing No. 1.
(*c*) a sectional elevation on the cutting plane S-S.
Show hidden detail in the plan only.
In a suitable title block at the bottom of the paper print:
(*a*) the title.
(*b*) the scale.

Also indicate the angle of projection used. All construction lines should be erased.

Middlesex Regional Examining Board

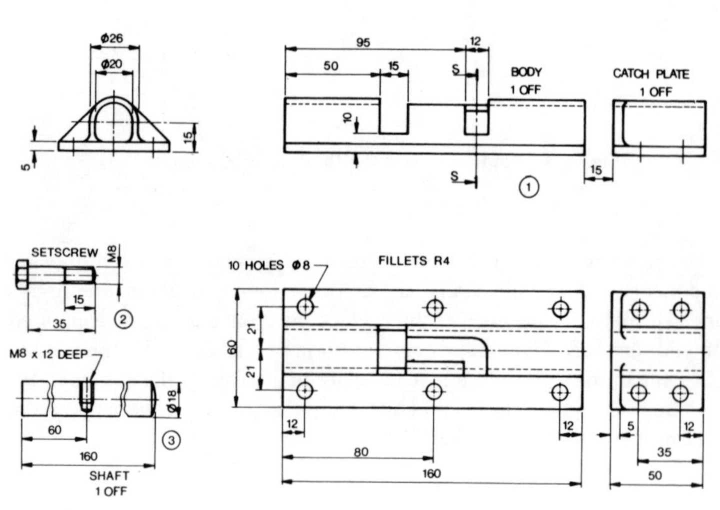

Fig. 7.45

Bevel gear mounting bracket

Fig. 7.46 shows a bevel gear mounting bracket.

(*a*) Draw, using first or third angle projection, the following:

 (i) the given plan showing hidden detail.

 (ii) a sectional elevation on plane A-A.

 (iii) an end elevation looking in the direction of arrow X and taken as a section on plane B-B.

(*b*) Add the following to your drawing:

 (i) a suitable title.

 (ii) scale.

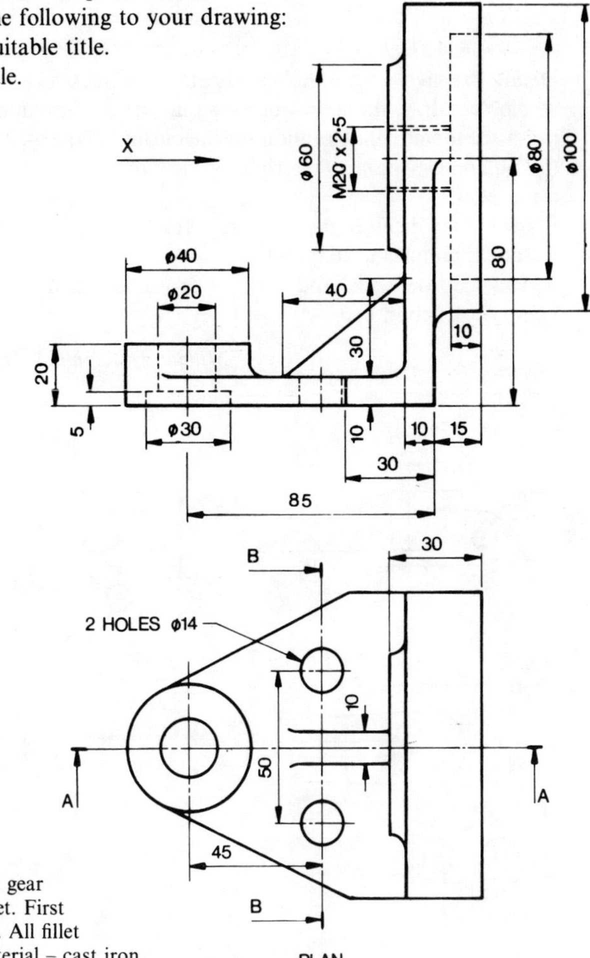

Fig. 7.46 Bevel gear mounting bracket. First angle projection. All fillet radii 4 mm. Material – cast iron.

(iii) the appropriate symbol to indicate the projection angle.

(iv) machining symbols on two surfaces which you would expect to be machined.

(v) dimension a counterbore, a screw thread, a centre distance, an outside diameter and a fillet radius.

City and Guilds of London Institute

5 *Speed control lock lever*

Details are given of the three parts of a lathe lock lever. With the parts assembled, draw the following views in first or third angle projection:

(*a*) a sectional front elevation on the cutting plane S-S,

(*b*) an end elevation to the right of view (*a*),

(*c*) a plan.

Show hidden detail in the plan only. Insert six dimensions, which must include a radius and diameter.

Add the title, scale, and angle of projection used. You should erase your construction lines.

Middlesex Regional Examining Board

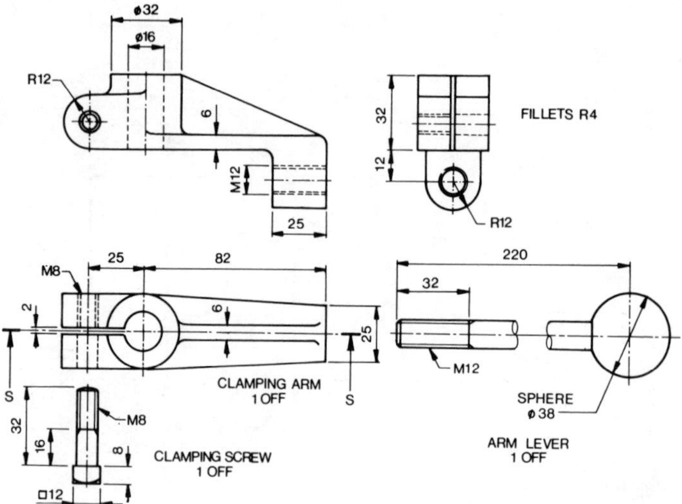

Fig. 7.47

6 *Lathe rest assembly*

The drawings below show the three components of a lathe rest assembly.

The tool rest (1) fits into the housing (2) and is held in position by the locking screw (3). With the three parts assembled so that face B of the tool rest is 8 mm above face C, draw full size, in first or third angle projection, the following views:

(*a*) a front elevation viewed in the direction of arrow A,
(*b*) a sectional end elevation on S-S,
(*c*) a plan.

Any dimensions not shown are left to your judgment. Do not show hidden detail.

Insert the following dimensions on the locking screw (3):

(*a*) size and type of thread.
(*b*) diameter of head.
(*c*) diameter of tommy bar.
(*d*) length of tommy bar.
(*e*) radius of end of head.

Add a title block containing: type of projection used, name of assembly and scale.

Middlesex Regional Examining Board

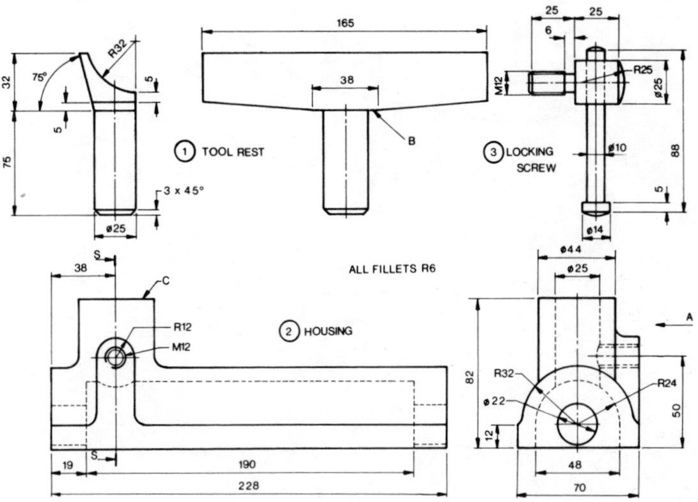

Fig. 7.48

7 *Drilling machine depth gauge*

The detail drawings show parts of a drilling machine depth gauge.

The M 12 screw fits into the M 12 hole of the body and is locked in position with the special locknut under the body. With the three parts assembled, the screw projecting 144 mm above the top face of the body, draw the following views full size, in first or third angle projection:

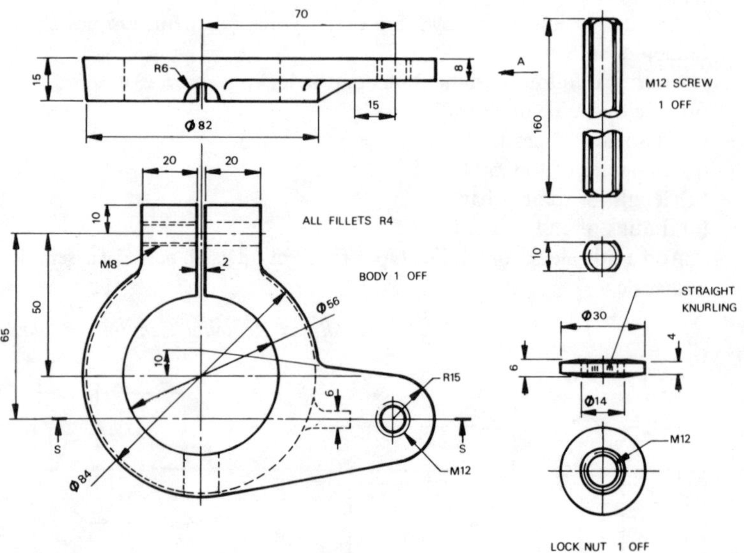

Fig. 7.49

(*a*) A sectional front elevation on the cutting plane SS.
(*b*) An end elevation viewed in the direction of arrow A.
(*c*) A plan showing hidden detail.

In the title block at the bottom right hand corner give the following information:

(*a*) The name of the assembly.
(*b*) The scale.
(*c*) The angle of projection used.

All construction lines should be erased.

Middlesex Regional Examining Board

Solutions

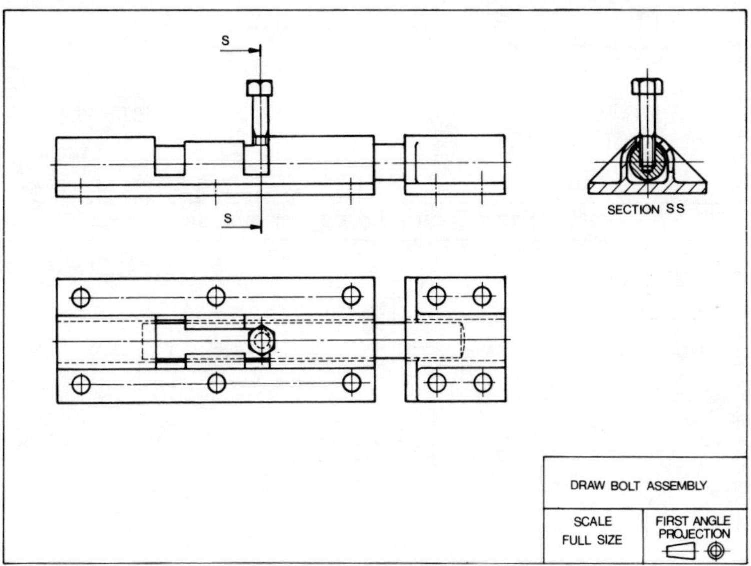

Draw Bolt solution

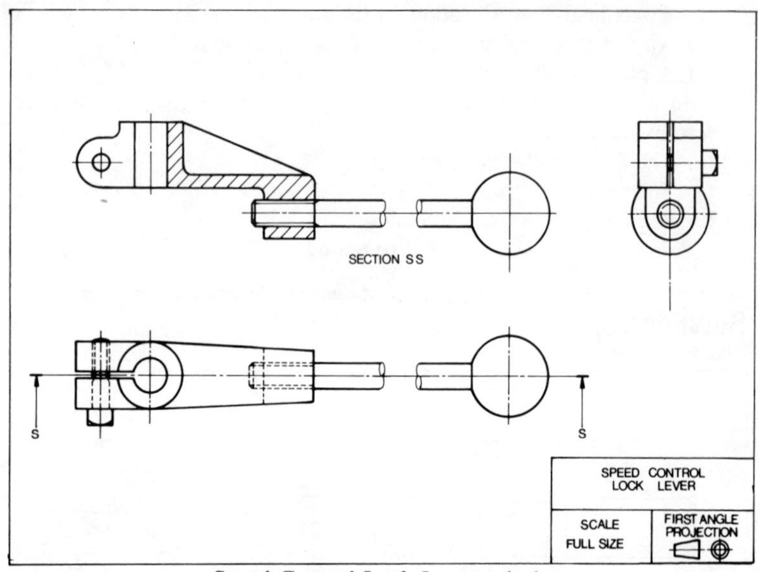

Speed Control Lock Lever solution

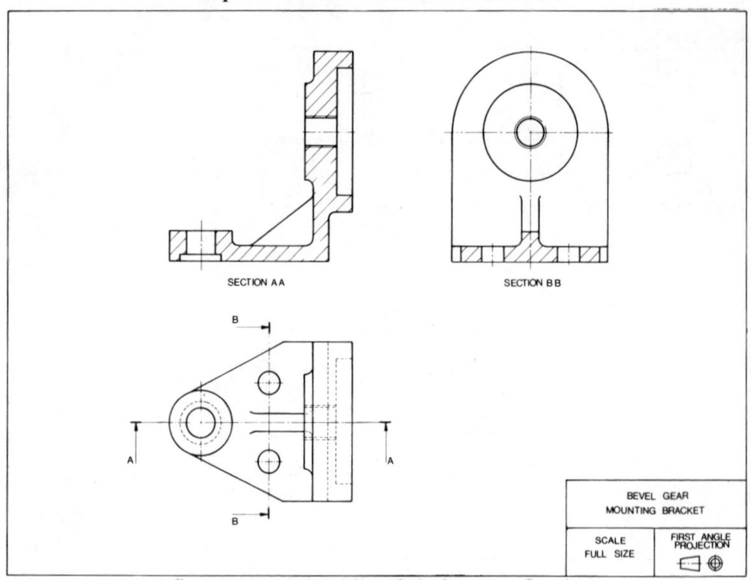

Bevel Gear Mounting Bracket solution

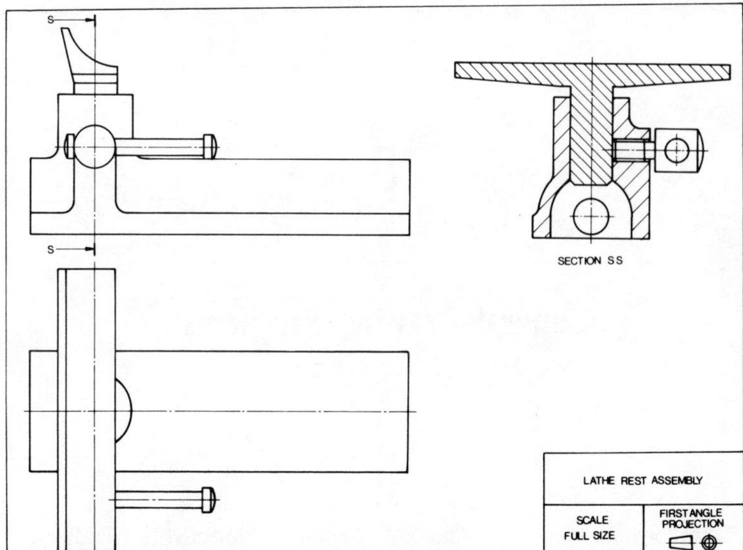

Lathe Rest Assembly solution

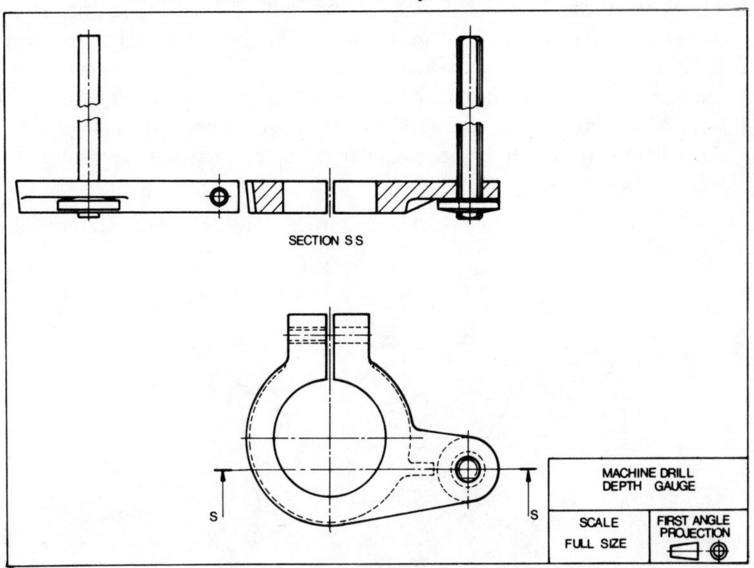

Drilling Machine Depth Gauge solution

8

General Drawing Problems

The following examination standard questions are printed to gain further drawing experience in all aspects of the subject previously described in this book. Work through the problems methodically. The additional practice will enable you to acquire speed and efficiency in technical draughtsmanship coupled with improved linework and general presentation technique. Take particular care with small details such as lettering, dimensioning, arrowheads, etc., and notice how the overall appearance of your drawings improve.

All answers are given at the end of the chapter so that your solutions can be checked.

1 Fig. 8.1 shows the ebony stock of a mortice gauge. The front is elliptical

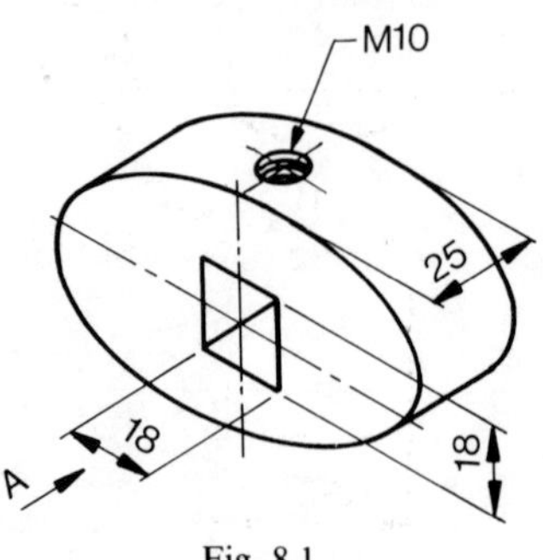

Fig. 8.1

in shape with the major axis 70 mm and the minor axis 50 mm; the M10 threaded hole is central in the top face. Draw a front elevation from A, and the plan of the stock, showing clearly your construction for the true ellipse. Show hidden detail on the front elevation

Middlesex Regional Examining Board

2 Fig. 8.2 shows a sectional elevation through a component which has been manufactured by the casting process. Name the features indicated by the letters E, F, G, H, K and M.

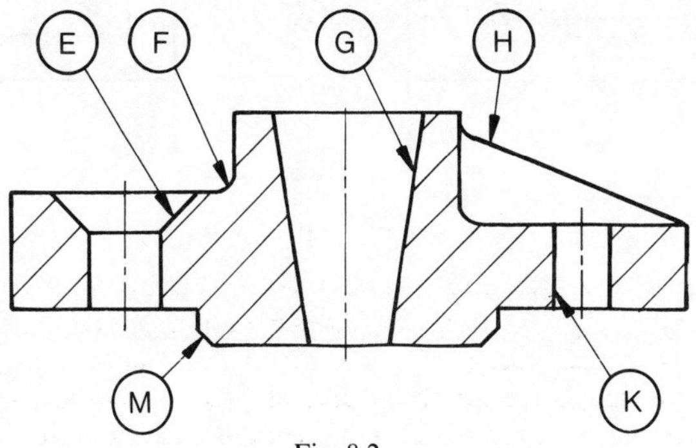

Fig. 8.2

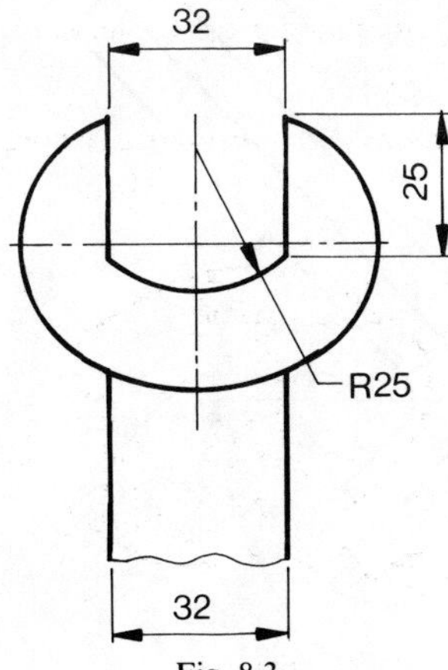

Fig. 8.3

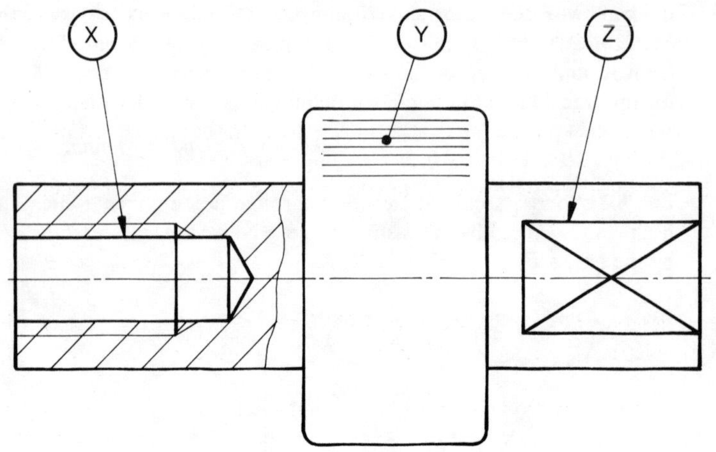

Fig. 8.4

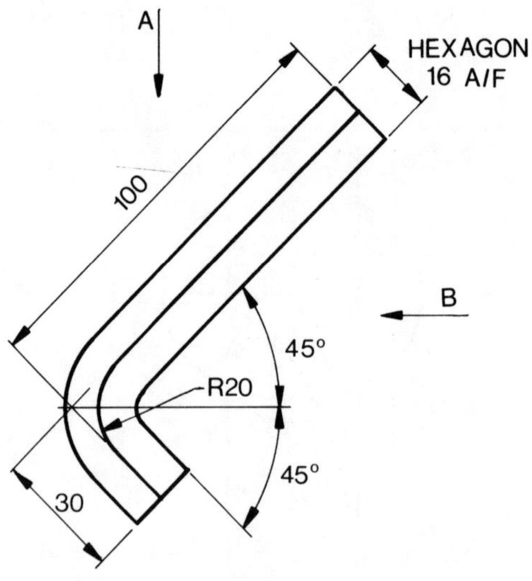

Fig. 8.5

3 Find by construction, and using Pythagoras' theorem, the square root of 2, 3, 4, 5, 6 and 7.

4 The open ended spanner is of elliptical form with the major axis 65 mm and the minor axis 50 mm. Draw the given view, full size, showing your construction for the ellipse and the arc centre.

In the open jaw, draw a regular hexagon 32 A/F to represent the outline of a suitable hexagonal nut.

Middlesex Regional Examining Board

5 The conventional representation of three different parts is shown by the letters X, Y and Z in Fig. 8.4. What do the letters indicate?

6 Fig. 8.5 shows the elevation of a key for a socket screw. The key is manufactured from hexagonal shaped steel bar. Project a plan and end elevation in the direction of arrows A and B in first angle orthographic projection.

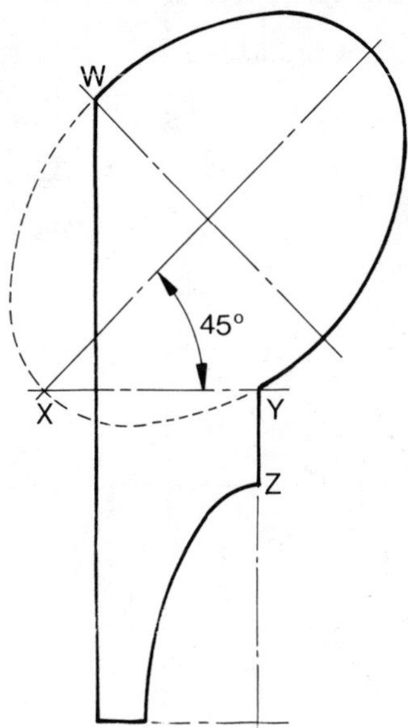

Fig. 8.6

7 A moulding suitable for a picture rail is seen from its section to be constructed largely from elliptical curves.
The larger ellipse has axes of 100 mm × 70 mm.
The smaller ellipse has axes of 100 mm × 50 mm.
The vertical back of the moulding is drawn from the end of the minor axis (W).

A horizontal line is drawn from where the major axis meets the ellipse to cut at X and Y. From Y a vertical is dropped for a distance of 20 mm to Z, and continues as the major axis of the smaller ellipse.

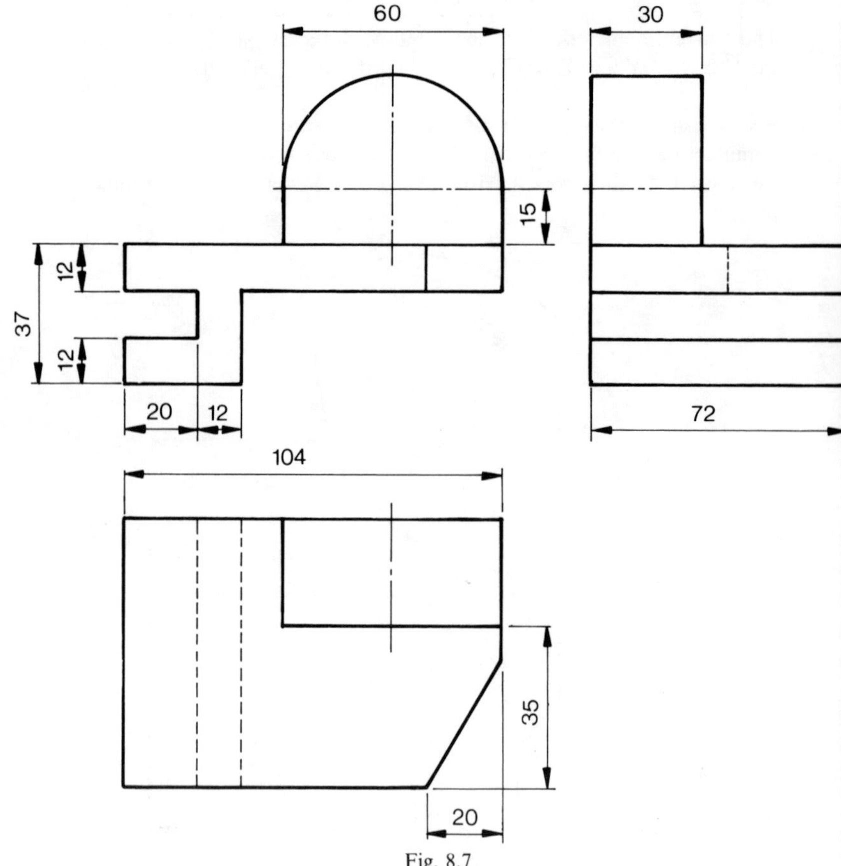

Fig. 8.7

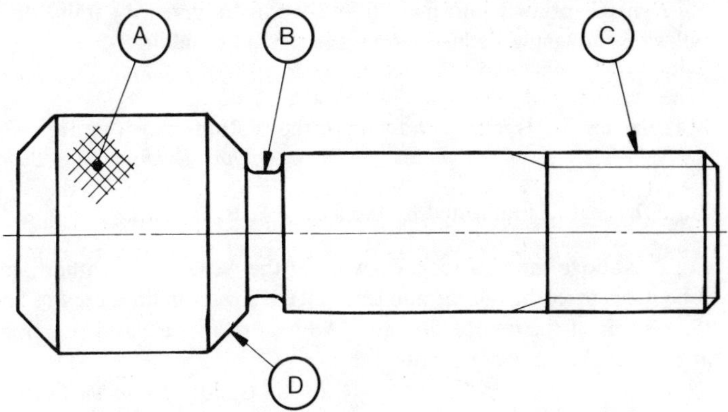

Fig. 8.8

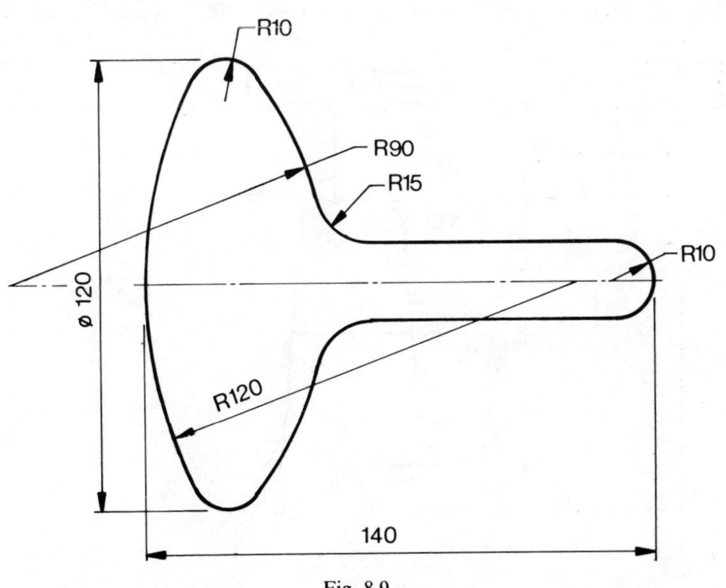

Fig. 8.9

From the printed information and from the given diagram, draw full size, showing all construction, the shaped moulding.

East Anglian Examinations Board

8 Make a full size isometric drawing of the bracket shown in Fig. 8.7.

East Anglian Examinations Board

9 State the features indicated by the letters A, B, C and D in Fig. 8.8.

10 Fig. 8.9 above shows a scale drawing of the outline of a cathode ray tube. Construct the outline and leave all construction lines used to find the centres of the intersecting arcs. Add short lines across the outline at *each* of the tangency points.

City and Guilds of London Institute

11 Fig. 8.10 shows the outline of a water tap. Reproduce the drawing full size, showing your construction for the arc centres and the points of tangency.

Middlesex Regional Examining Board

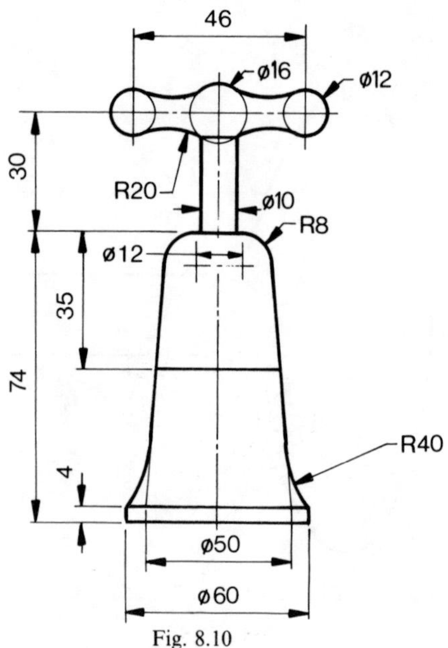

Fig. 8.10

12 The profile of a reclining seat bracket is shown in Fig. 8.11. Redraw the given view, full size, to the dimensions given, showing clearly the construction of all tangents and tangential arcs. Points of tangency should be indicated by a short line across the outline.

East Anglian Examinations Board

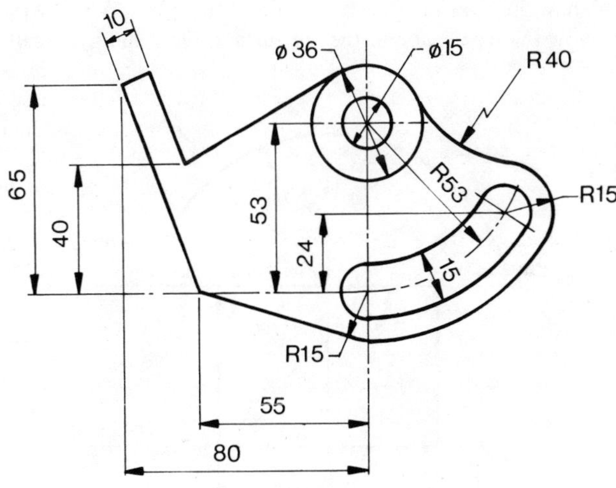

Fig. 8.11

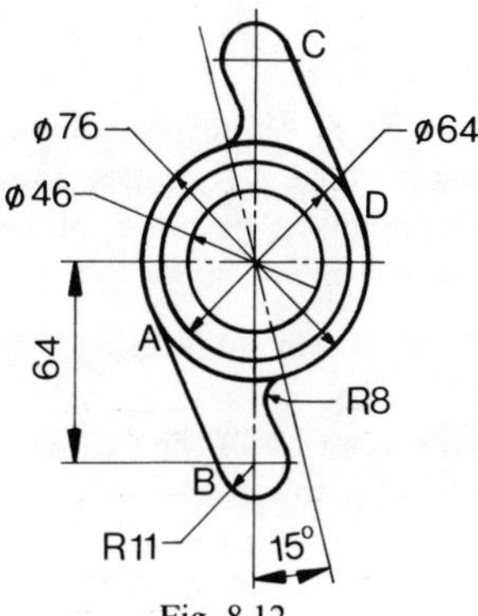

Fig. 8.12

13 Fig. 8.12 shows a 'knock on' hub as fitted to a sports car. Draw the given figure, full size, showing your construction for the tangents AB and CD and the arc centres.

Middlesex Regional Examining Board

14 A short rivet is shown in Fig. 8.13. Copy the given view and project a plan of the rivet which is below the cutting plane AA. Project a view to show the true shape of the cut surface. Omit hidden detail.

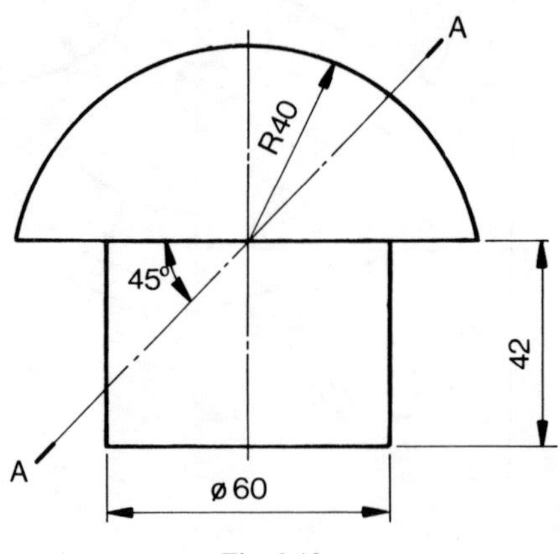

Fig. 8.13

15 The pictorial sketch (Fig. 8.14) shows an inlet fitting on a storage tank. The fitting is in the form of a right hexagonal prism with a central circular hole. From the given elevation, project a plan of the fitting. All hidden detail should be indicated.

North Western Secondary School Examination Board

16 Make a full isometric drawing of the casting in Fig. 8.15. No hidden detail or dimensions are required. Start your drawing with the point X in the foreground.

East Anglian Examinations Board

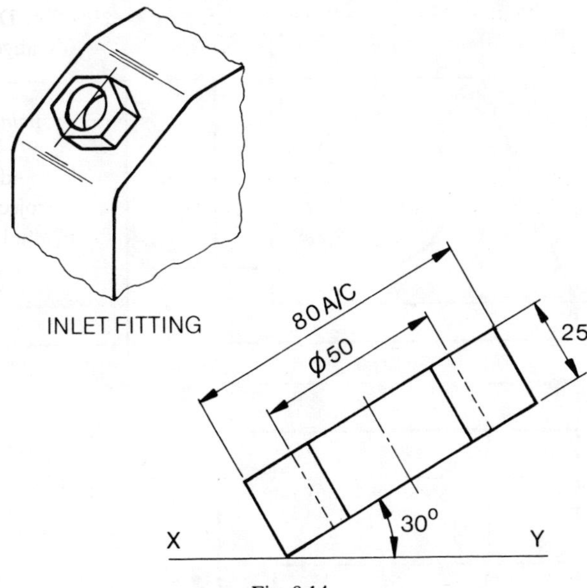

Fig. 8.14

17 Fig. 8.16 shows part of an air duct which is manufactured from thin sheet metal.

(*a*) Calculate the length of the developed surface.

(*b*) Divide the developed surface length into twelve equal parts using a geometrical construction.

(*c*) Plot the development of the curved surface.

Make no allowance for thickness or overlap at the seam.

City and Guilds of London Institute

18 A metal disc is 140 mm in diameter and 30 mm in thickness, as shown in Fig. 8.17 below. The outside edge of one side of the disc is chamferred at 45° and the depth of the chamfer is 5 mm. In the centre of the disc is a hole which has been tapped with a 20 mm thread. The pitch of the thread is 2·5 mm. Three holes, 12 mm diameter on a 100 mm pitch circle diameter are also drilled through the disc and these holes are

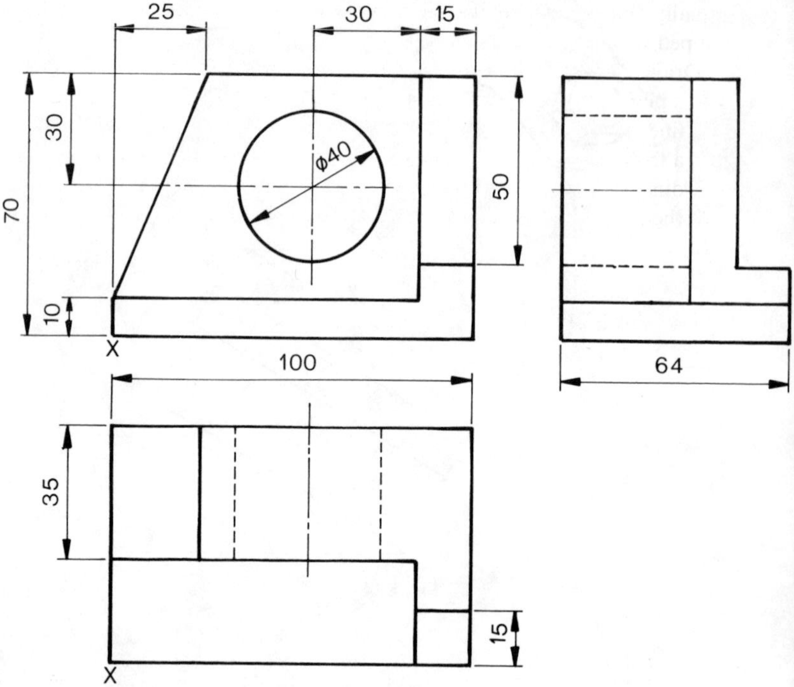

Fig. 8.15

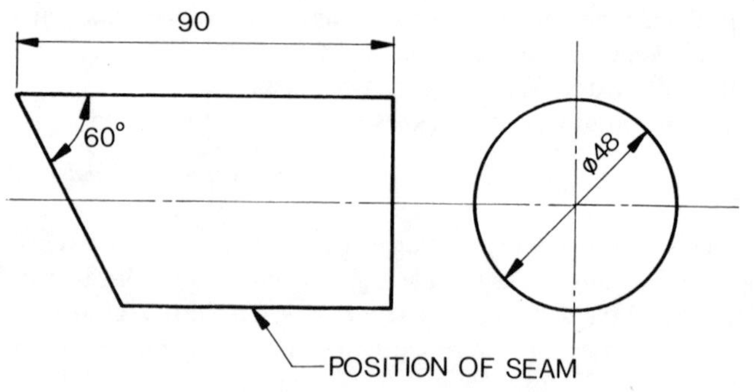

Fig. 8.16

equally spaced around the pitch circle which is concentric with the tapped hole.

Draw full size, and in third angle projection the following:

(*a*) a plan with one of the 12 mm holes drawn on the horizontal centre-line.

(*b*) a front elevation taken as a section as shown below in the outline.

(*c*) dimension your solution correctly to British Standard 308.

(*d*) the projection symbol for third angle projection.

City and Guilds of London Institute

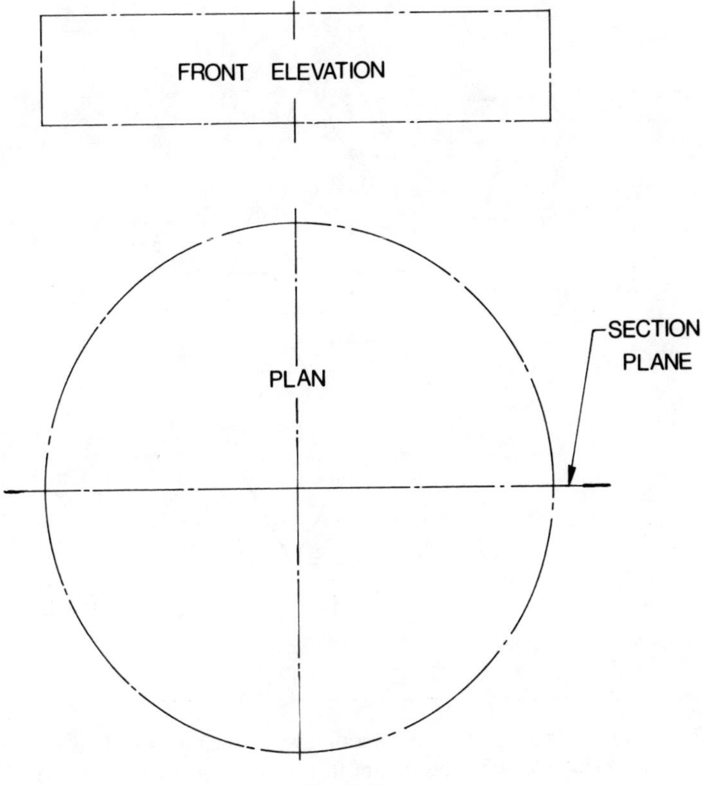

Fig. 8.17

19 Fig. 8.18 below shows a solid component which has been cut from a hexagonal pyramid. Draw, full size and in first angle projection, the following views of the solid.

(*a*) the given front elevation.
(*b*) complete the partially drawn plan view.
(*c*) project an end elevation in the direction of arrow A.

City and Guilds of London Institute

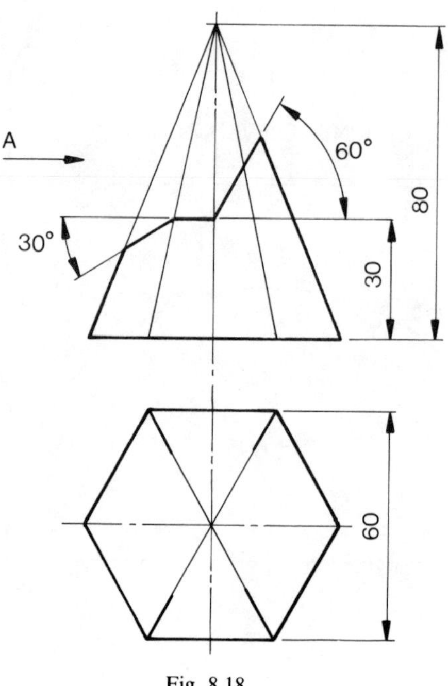

Fig. 8.18

20 An advertising sign consists of a thin wire which is wound into a right hand helix. Coloured beads are threaded along the wire so that the helix is divided into twelve equal sections. Part of the front elevation and plan view are shown in Fig. 8.19 with three beads in position. Copy the given views, construct the helix and show the remainder of the beads, correctly located, in both views.

City and Guilds of London Institute

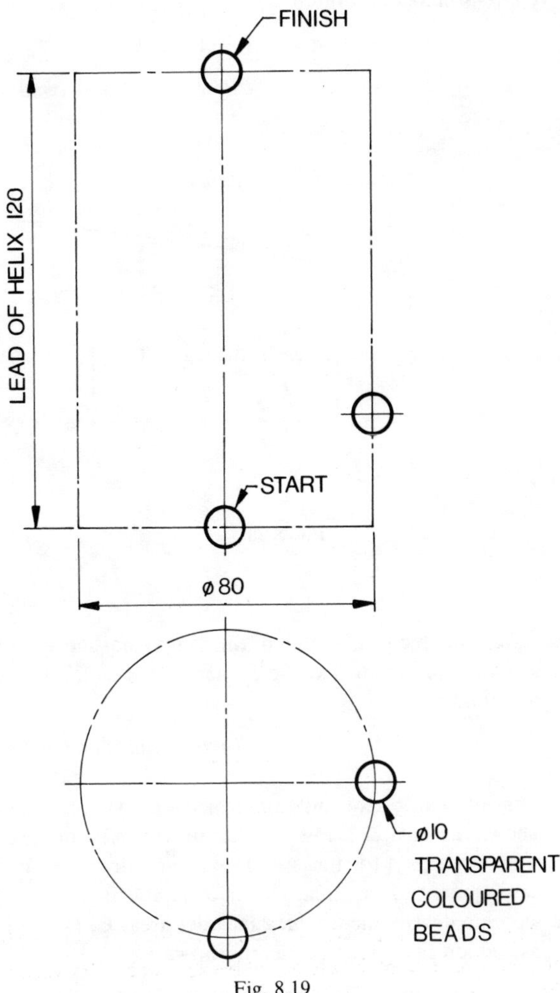

Fig. 8.19

21 A line diagram of a simple mechanism is shown in Fig. 8.20 below where crank OA rotates in an anticlockwise direction about centre O. The connecting rod AB slides through a pivot which is free to rotate about centre C. Plot, full size, the locus of point B for one complete revolution of the crank OA. In your construction choose suitable points to give a reasonably accurate locus.

$$OA = 35 \text{ mm}, AB = 140 \text{ mm}$$

City and Guilds of London Institute

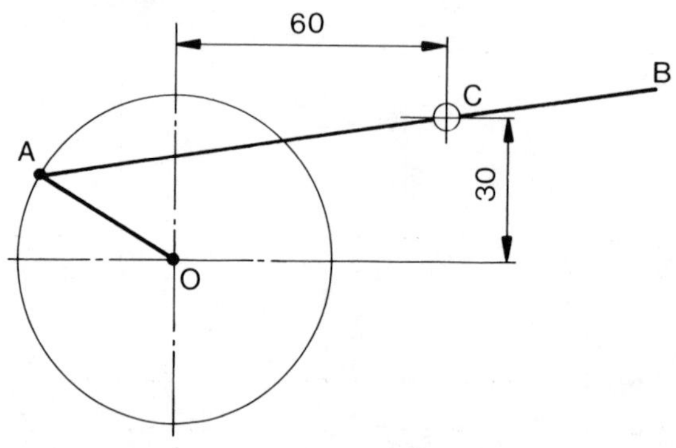

Fig. 8.20

22 Make a freehand pictorial sketch of the component shown in Fig. 8.21. Present your solution so that the corner shown as 'X' will appear in the foreground.

City and Guilds of London Institute

23 The outline of a cardboard container which is used to package an Easter Egg is shown in Fig. 8.22 below. Neglect the thickness of the cardboard and the material used for tucks and joints and draw the following:
(*a*) a development of the areas shown as A and B.
(*b*) a separate development which includes areas C, D and E.
All construction lines must be clearly shown.

City and Guilds of London Institute

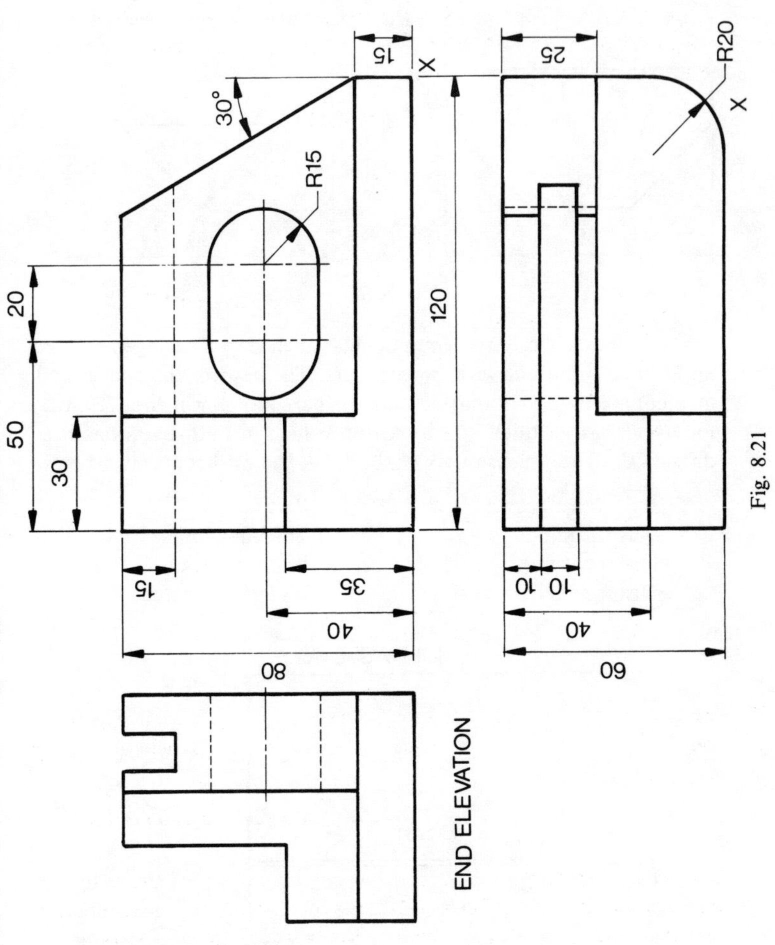
15
X
30°
R15
20
50
30
120
15
35
40
80
END ELEVATION
25
R20
X
10
10
40
60

Fig. 8.21

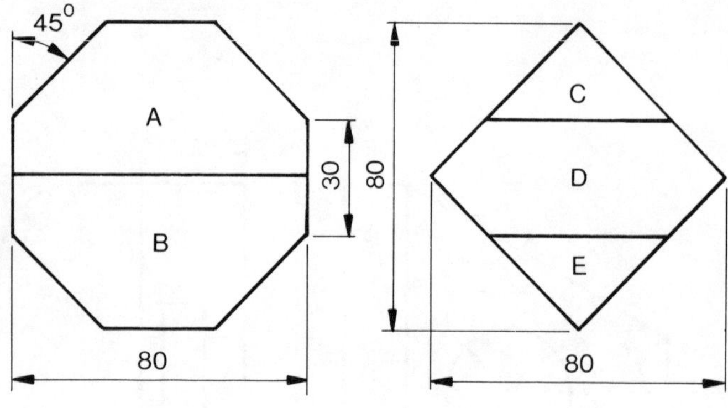

Fig. 8.22

24 Fig. 8.23 shows part of two components which are bolted together using an M30 bolt and nut with two washers. The head of the bolt is sunk in a counterbore Ø70 and 30 deep in part No. 1 which is 50 thick. Part No. 2 is 35 thick and a clearance hole in both parts has been drilled Ø32. The dimensions of the bolt, nut and washer are given below.

Bolthead thickness	19	Bolt length	95
Nut thickness	24	Thread length	45
Distance across corners	53	Washer thickness	4
Distance across flats	46	Washer diameter	56

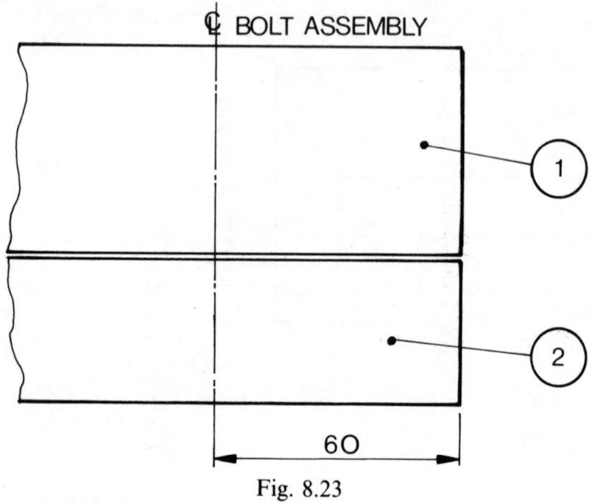

Fig. 8.23

(*a*) Draw a sectional elevation of the assembly.
(*b*) Project a plan view beneath the elevation using first angle orthographic projection and showing the counterbored hole, bolthead and washer only.

City and Guilds of London Institute

25 The component parts of a screw-jack for a test rig are given in Fig. 8.24. When assembled, the collar is fitted over the top of the jack screw and the part of the screw shown as 'X' is peened over so that the collar is still free to rotate. The screw is turned by the bar which is inserted and positioned centrally through the 4 mm hole in the jack screw. The screw rotates in the jack base.

Make an assembly drawing to a scale of 1 : 1 in first or third angle projection, and showing no hidden detail. The following views are to be shown

(*a*) a sectional elevation with the bar positioned with its centre 85 mm above the bottom edge of the base.
(*b*) an outside end view.
(*c*) an outside plan view.

Add reference balloons to your solution to indicate the components and the appropriate projection symbol.

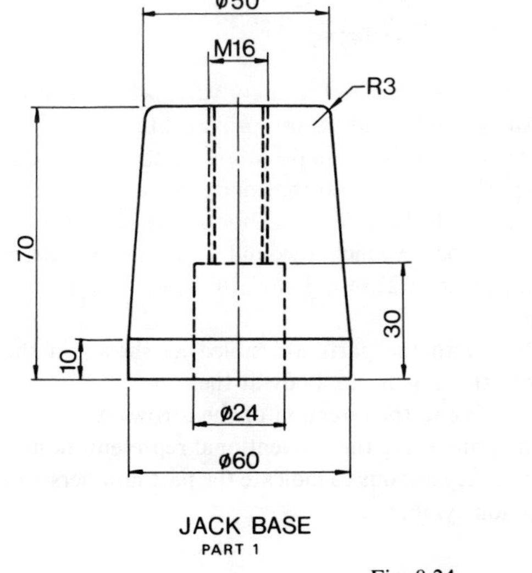

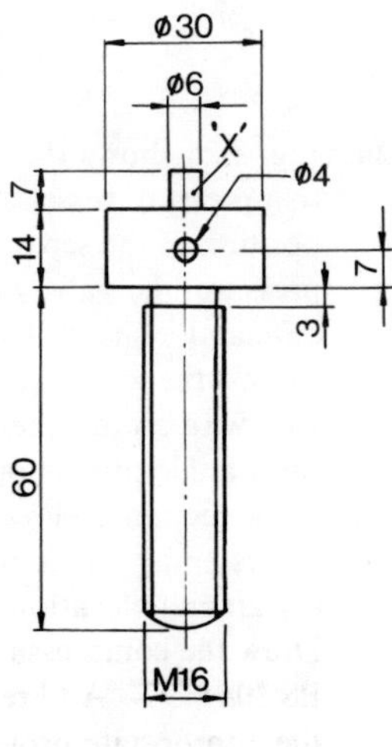

Fig. 8.24

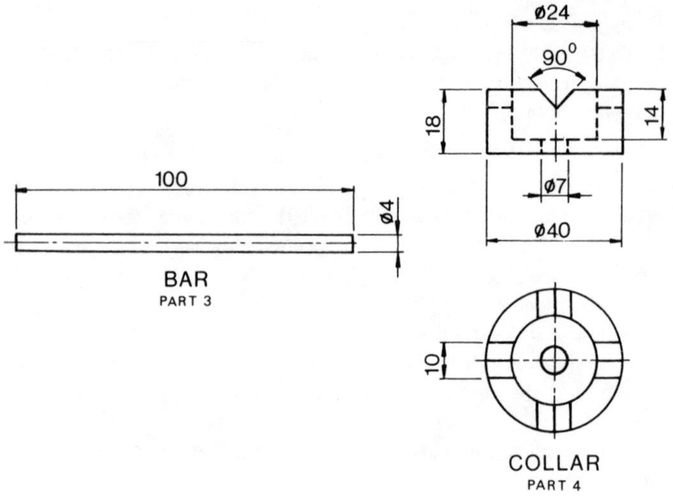

Fig. 8.24 (continued)

26 Fig. 8.25 shows the component parts of a safety valve which is fitted to a pressure vessel. An assembly guide is also provided to indicate the position of the separate parts. If the air in the vessel exceeds the design pressure, the valve is pushed away from the conical seat in the valve base and air is vented through the hole in the body. The valve is held against the seat by a spring, part 3 which is wound from 8 mm diameter wire with an outside diameter of 22 mm. Draw, full size, using first or third angle projection.

(*a*) a sectional elevation with the parts assembled as shown in the assembly guide with the hole in the body at the top.

(*b*) an end elevation looking in the direction of the arrow 'A'.

Draw the compression spring using the conventional representation in BS 308 : 1972. Add reference balloons to indicate the part numbers and the appropriate projection symbol.

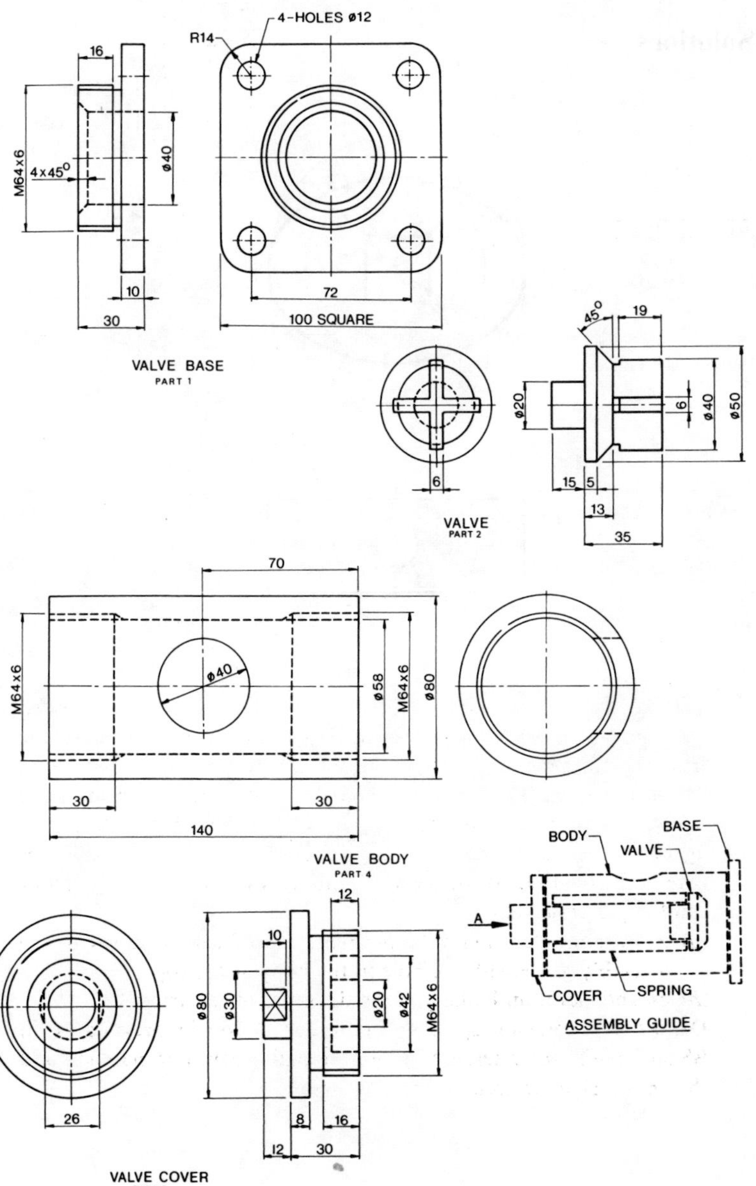
4-HOLES ø12
R14
16
M64x6
4x45°
ø40
10
30
72
100 SQUARE
VALVE BASE
PART 1
45°
19
ø20
6
ø40
ø50
6
15
5
13
35
VALVE
PART 2
70
M64x6
ø40
ø58
M64x6
ø80
30
30
140
VALVE BODY
PART 4
BODY
BASE
VALVE
A
COVER
SPRING
ASSEMBLY GUIDE
12
10
ø80
ø30
ø20
ø42
M64x6
26
8
16
12
30
VALVE COVER
PART 5

Fig. 8.25

Solutions

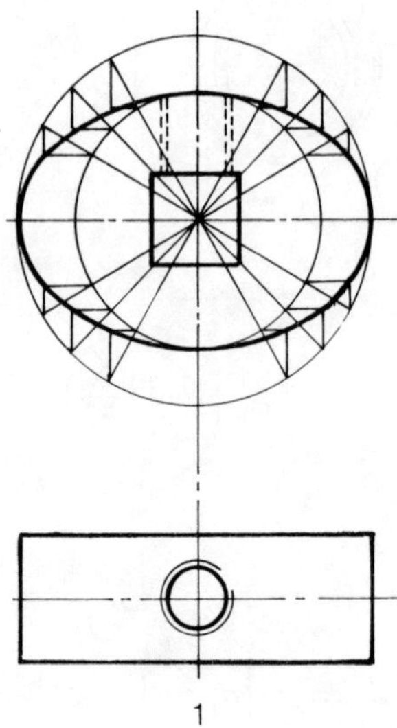

1

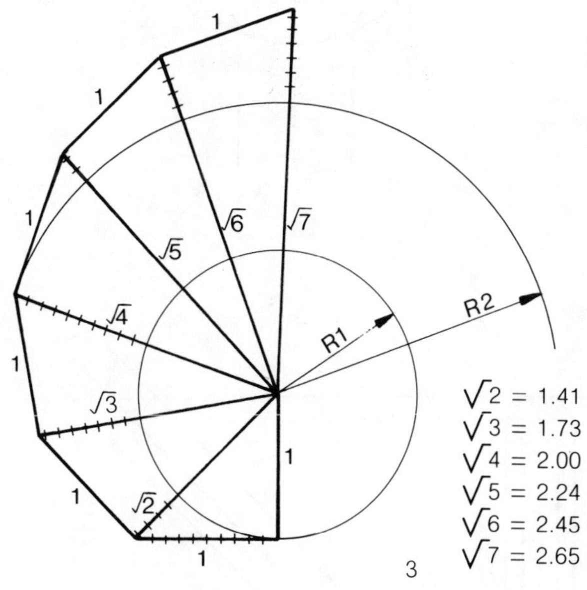

QUESTION 2 E — COUNTERSUNK HOLE, F — RADIUS,
G — TAPERED HOLE, H — WEB, K — HOLE,
M — CHAMFER

QUESTION 5 X — TAPPED HOLE OR INTERNAL THREAD,
Y — STRAIGHT KNURLING, Z — SPANNER FLATS

QUESTION 9 A — CROSS OR DIAMOND KNURLING,
B — UNDERCUT, C — EXTERNAL OR MALE THREAD

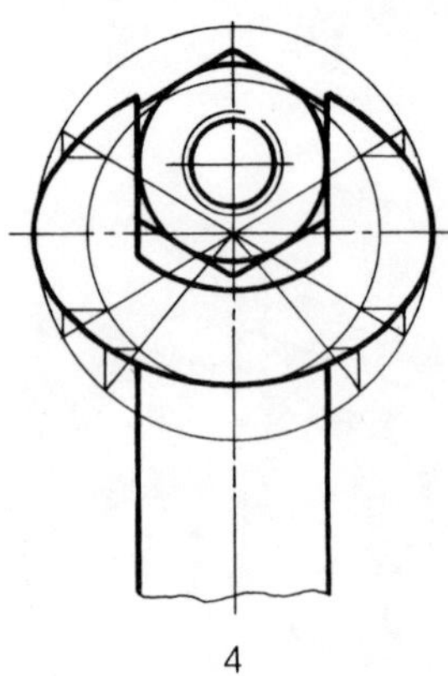

4

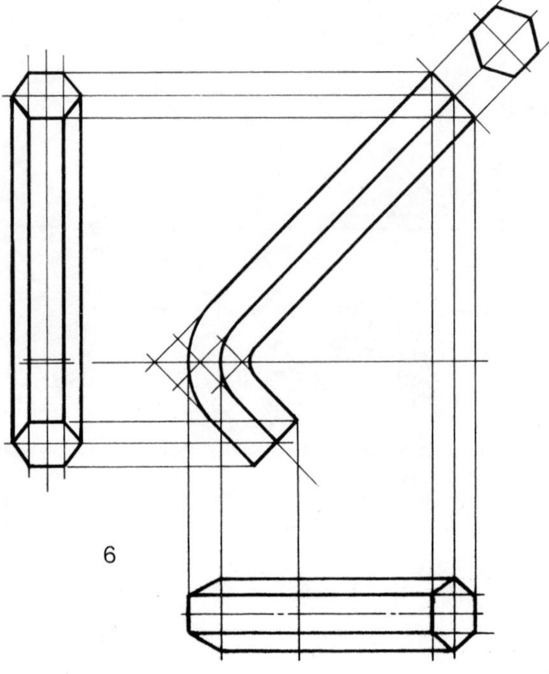

6

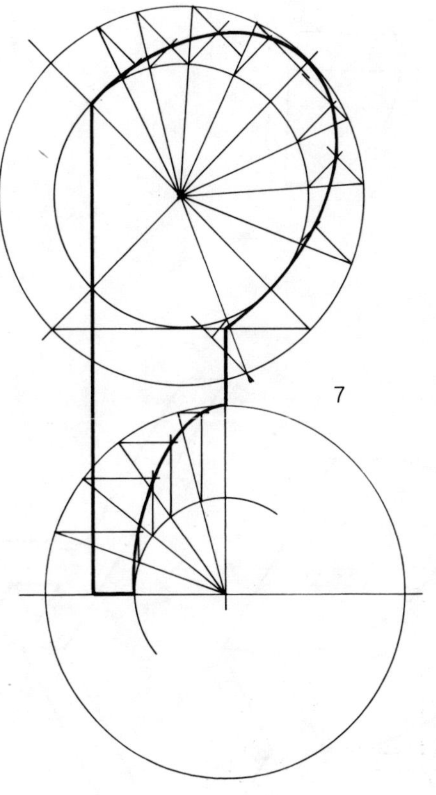
7

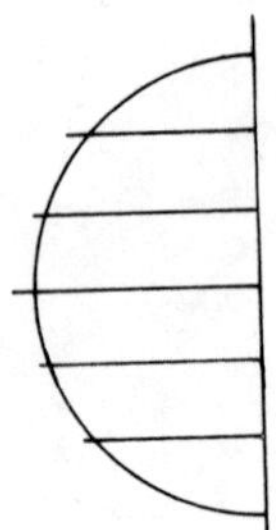

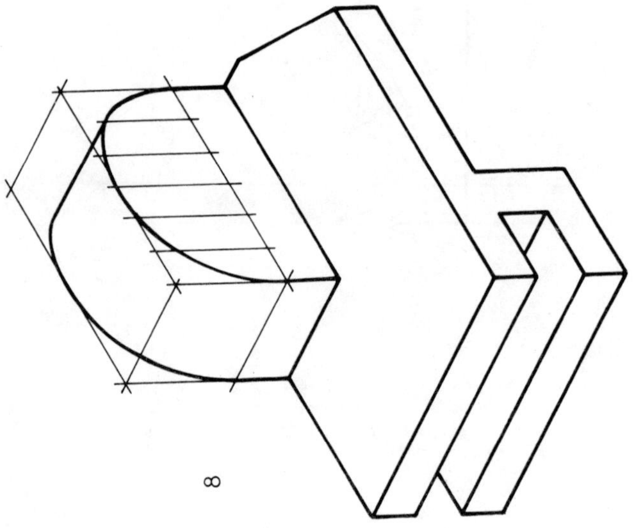

8

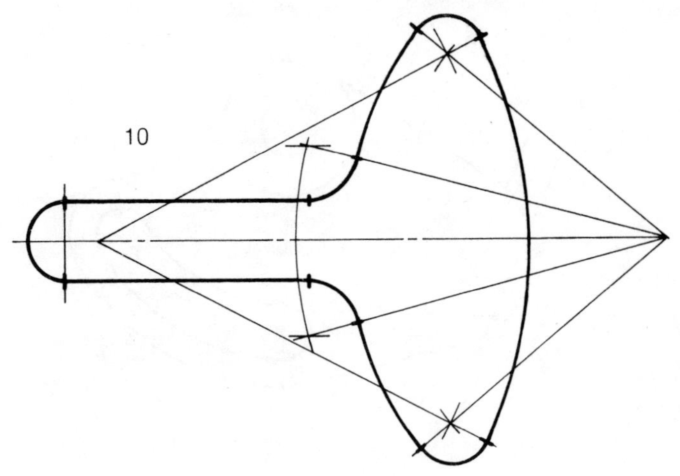

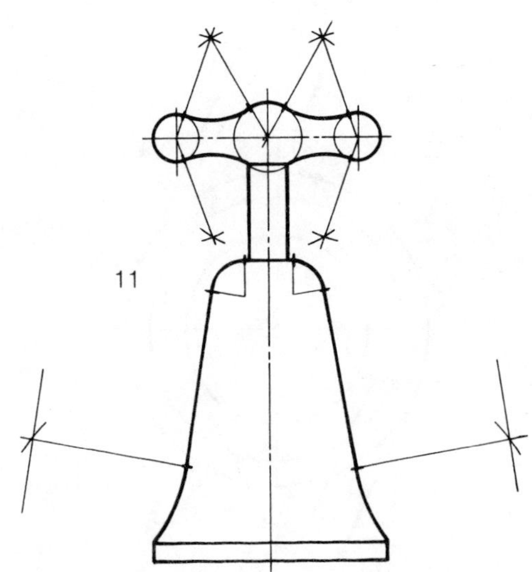

TANGENCY POINTS ARE SHOWN
BY DASHES ACROSS THE OUTLINE

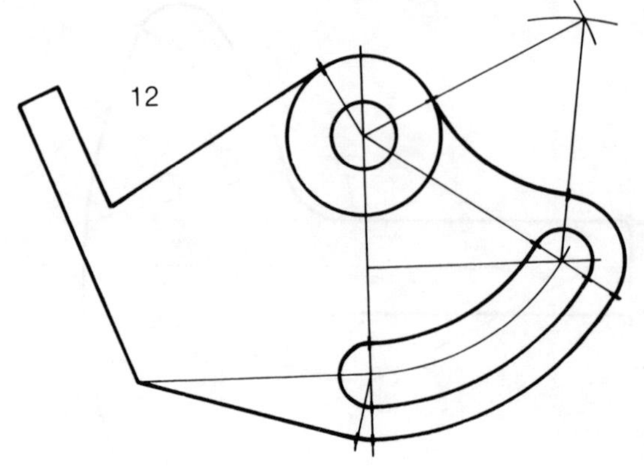
12

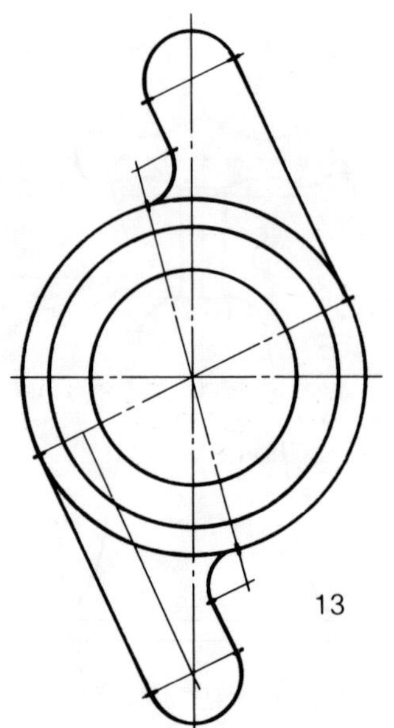
13

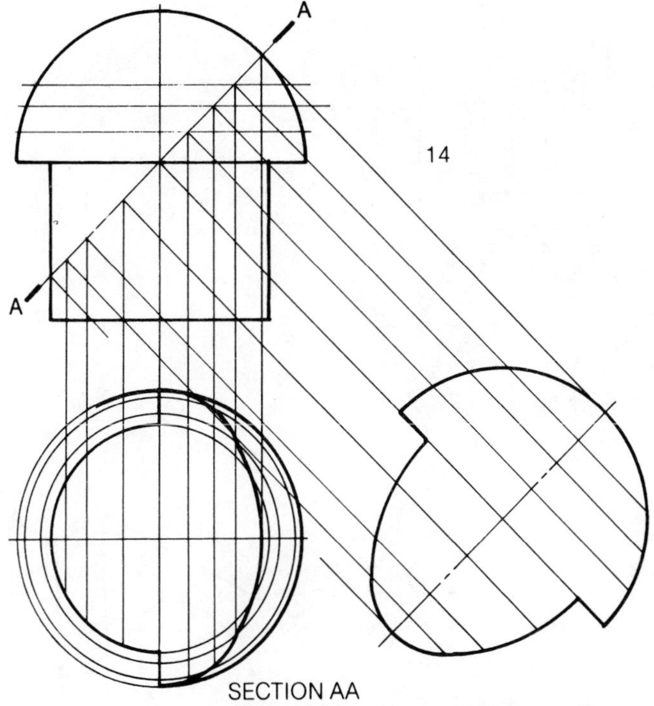

SECTION AA

NOTE THAT CROSS HATCHING LINES ARE OMITTED TO SHOW CONSTRUCTION CLEARLY

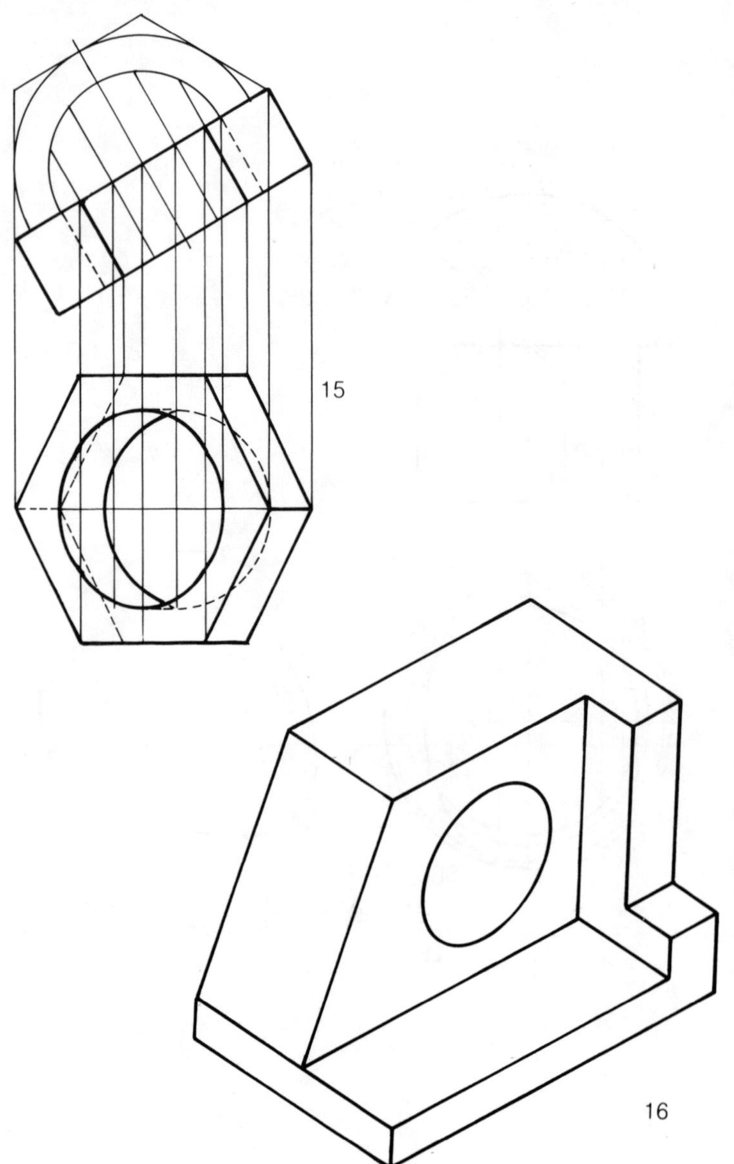
15
16

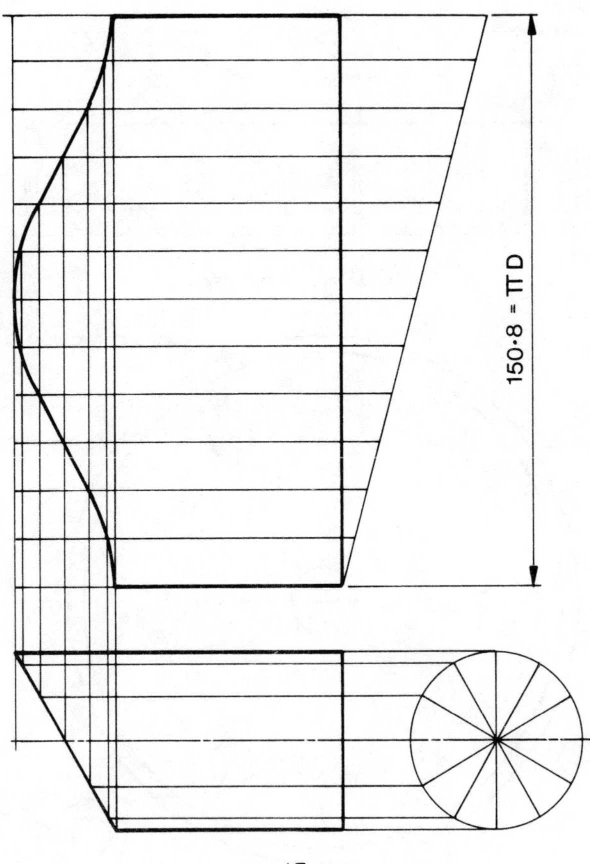

17

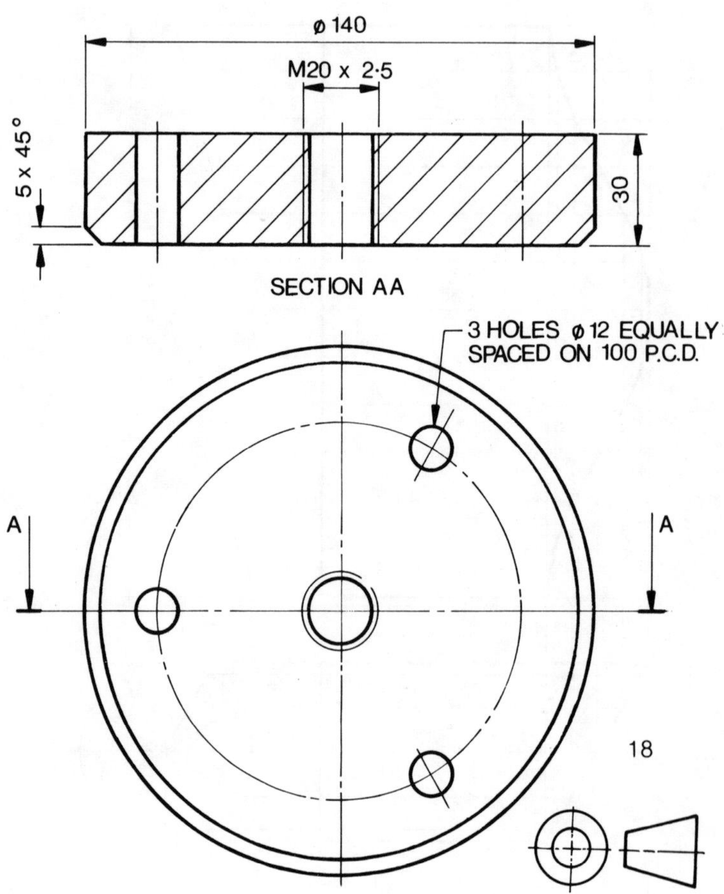

ø 140
M20 x 2·5
5 x 45°
30
SECTION AA
3 HOLES ø 12 EQUALLY SPACED ON 100 P.C.D.
A
A
18

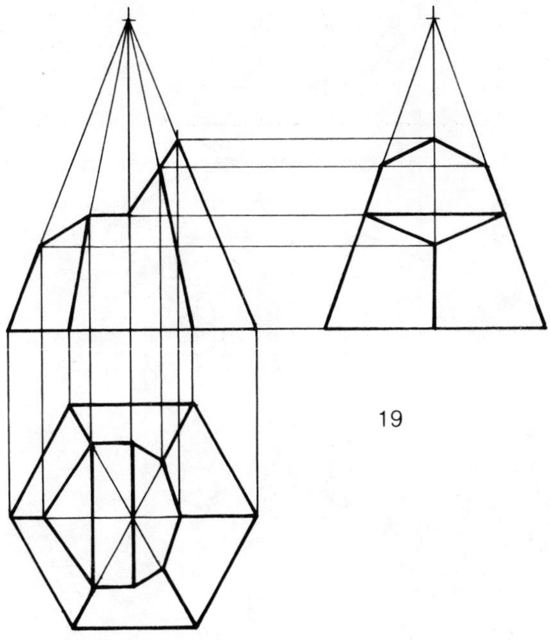
19

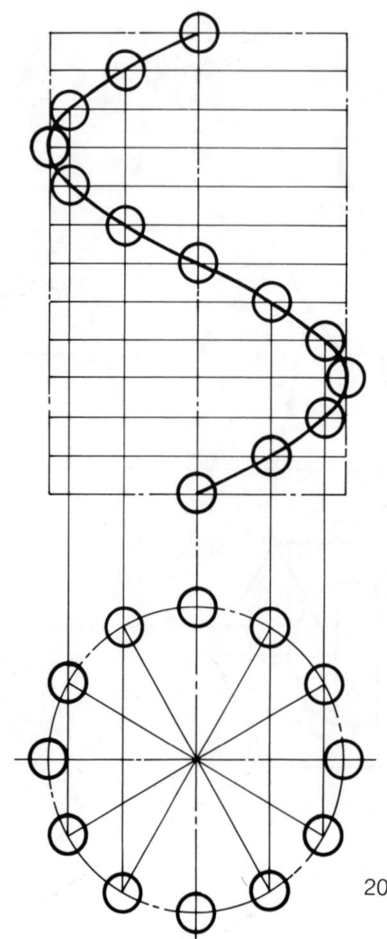

20

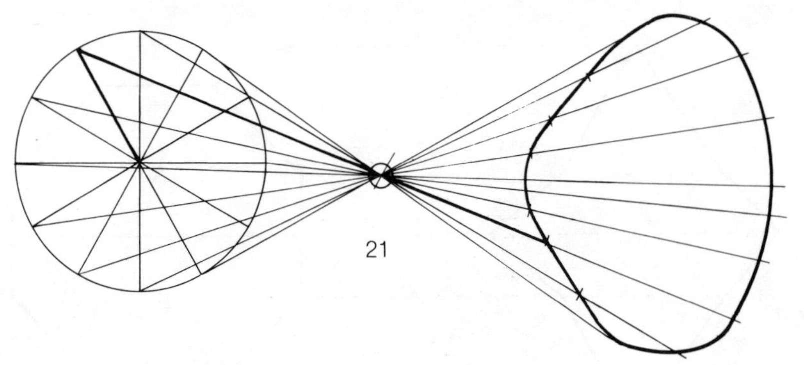

21

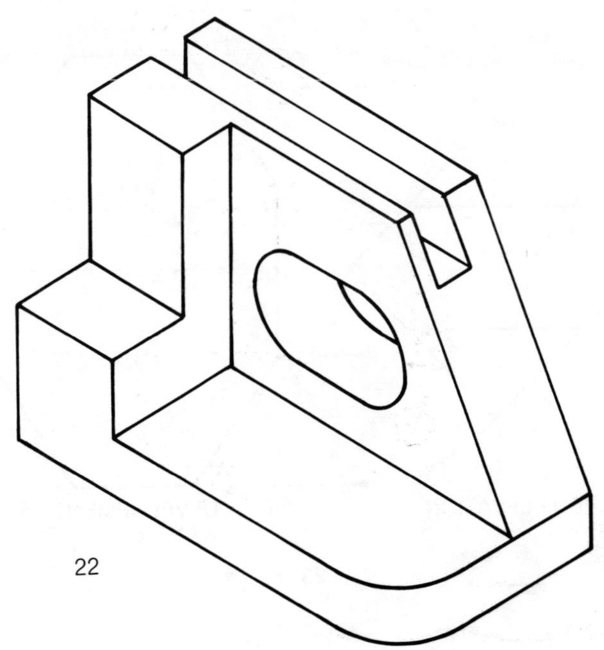

22

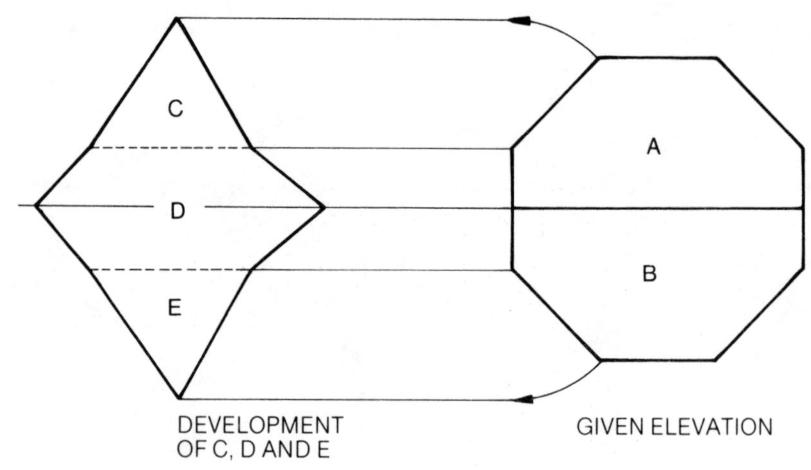
C
D
E
A
B
DEVELOPMENT OF C, D AND E
GIVEN ELEVATION

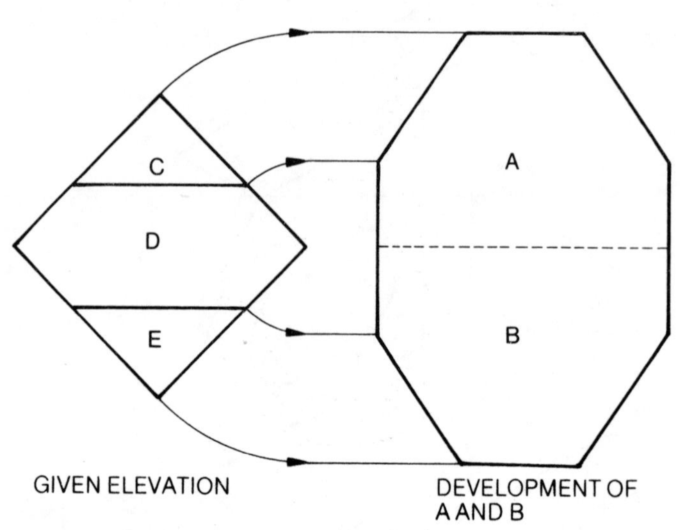
C
D
E
A
B
GIVEN ELEVATION
DEVELOPMENT OF A AND B

23

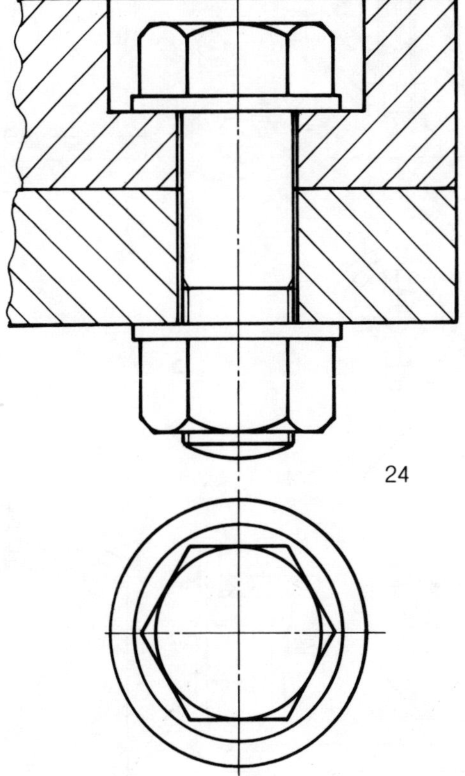
24

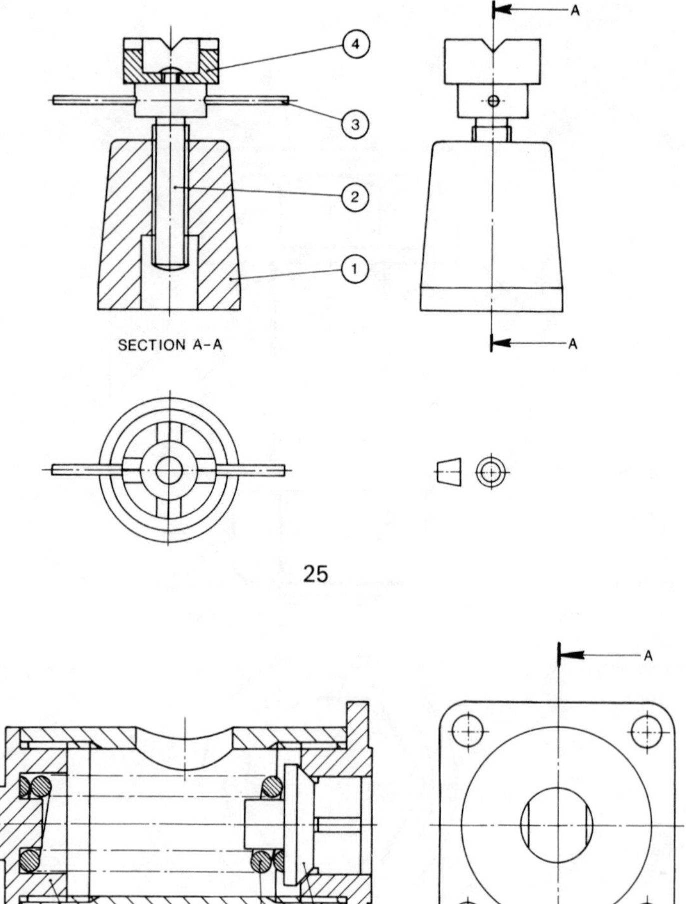

25

5 4 3 2 1

SECTION A-A

26

Index

GEOMETRY

P. ABBOTT

A knowledge of Geometry is basic to many disciplines, from engineering to architecture, and this clear introductory course guides the reader confidently through the necessary groundwork.

The text gives step-by-step explanations of both practical and abstract aspects of the subject, with numerous clear illustrations and exercises. A practical working textbook for use at all levels, in schools and at home.

TEACH YOURSELF BOOKS

MATHEMATICS

L. C. PASCOE

This step-by-step introduction offers clear explanations and numerous worked examples that will guide the reader to an understanding of essential mathematical concepts and techniques.

Beginning with a brief historical outline of the development of mathematics, the book gently steers the reader through the basics of arithmetical processes, algebra and geometry. In keeping with the times, it then focuses on the electronic calculator both as a computational aid in financial calculations and other practical applications, and as a useful tool in progressing to more advanced mathematics. Thus problems involving percentages, profit and loss, and interest calculations are explored – and trigonometry introduced – using the functions available on most modern calculators. Throughout, exercises (with answers) are provided to test and reinforce the reader's understanding.

TEACH YOURSELF BOOKS

PHYSICS

DAVID BRYANT

A complete and unified guide to elementary physics for the beginning student and interested layman who requires a modern introduction.

This book covers all aspects of elementary physics and includes lucid discussions of molecular and atomic structure, forces, energy and waves, and of the behaviour of light, gases and electricity. The text is illustrated with numerous sketches and photographs, and no previous knowledge is assumed on the part of the reader beyond a familiarity with basic mathematics.

'An exceptionally readable account of physics up to 'O' level standard.'

The Times Educational Supplement

TEACH YOURSELF BOOKS

THE POCKET CALCULATOR

L. R. CARTER and E. HUZAN

A practical handbook to help you master the techniques of using a pocket calculator.

This book explains calculator applications ranging from the basic arithmetical functions to a wide variety of mathematical, scientific, financial and statistical problems. As such, it is a book both for the beginner and for those wishing to make full use of the facilities available and get the most out of their calculator.

The techniques described are applicable to pocket and programmable calculators alike, and are fully illustrated with examples and exercises (with answers) at varying levels of difficulty.

TEACH YOURSELF BOOKS